Basic Construction Safety and Health
Fred Fanning

Copyright © 2014 Fred Fanning
All rights reserved.
ISBN-13: 978-1492982210
ISBN-10: 1492982210
Library of Congress Control Number: 2021901647

DEDICATION

To my uncle, Edgar Sylvester Hellyer

CONTENTS

Introduction

In this modern age, it is a shame that we still have thousands of workers injured or made ill by their work. With all the modern appliances, medicine, technology, and knowledge it is completely unacceptable for workers to suffer from work. In the mid-1970s, our society decided that employers have a responsibility to send workers home in as good a condition as they arrived. Since that time there is progress, but not enough. One area that needs more improvement is the construction industry.

Construction is a high-hazard occupation because it exposes workers to heights; heavy equipment; unguarded machinery, power, hazardous materials; and trenches. Working with these hazards requires employers to implement effective safety and health programs. My focus in this book is on the four main risks for construction including, struck-by, caught in or between, and electrocutions. Furthermore, I cover basic safety topics for stairways and ladders, electricity, fall protection, tools, welding and cutting, personal protective equipment (PPE), motor vehicles, and excavations.

Aside from the hazards I also address a couple of unique circumstances that exist with construction sites. First, construction sites involve contractors and subcontractors and are multiple employer sites. Second, the Public Access Doctrine allows an Occupational Safety and Health Administration (OSHA) compliance officer to cite an employer without gaining access to the site. Citing this way is allowed because construction sites are often large, and compliance officers can stand on public property, identify risks, and cite the employer without ever gaining access to the site.

Effective management of worker safety and health programs reduces and eliminates the hazards that construction workers are exposed to. Effective construction safety and health programs start with a written document that outlines management and employee responsibilities, worksite analysis, hazard prevention and control, and safety and health training.

In exemplary companies' senior managers assign responsibilities to managers, supervisors, and employees so that everyone is involved in preventing accidents. Fulfilling these responsibilities leads to hiring the right workers, orienting, and training them, inspecting jobs on a regular basis, reporting, and investigating accidents, and enforcing safety and health standards at all levels.

Reducing or eliminating hazards leads to reductions in the extent and severity of work-related injuries and illnesses. Reducing injuries and illnesses can improve employee morale and productivity as well as reduce the cost of:

- worker compensation claims
- damage to equipment
- insurance premiums
- lost productivity

To help you make sense of all this information, I cite sources from publications and organizations from which the information was gathered. It is essential that you be able to refer to the original sources when necessary. I recommend that anyone working in construction become familiar with 29 CFR Part 1926 and applicable sections of 29 CFR Part 1904 and Part 1910.

I can only cover so much in this short book. For more in-depth information, I recommend two books. The first book is "Construction Safety Handbook" 2nd edition, by Mark Moran, published by Government Institutes, Inc. in 2003. The second book is "Construction Safety Management and Engineering" 2nd edition, by Darryl C. Hill, Ph.D., CSP, the editor published by the American Society of Safety Engineers.

In addition to these references, you can gain further insight by attending either a 10- or 30-hour OSHA construction safety course. Either course teaches the necessary information you or those that work for you need to prevent becoming the victim of a work-related accident or illness.

There are OSHA publications that provide pithy information on subjects. In the chapters, I include specific information about these publications to help you obtain them. I hope you go to osha.gov and check them out.

I spent 10 years working in construction before I became a safety professional. I then spent over 20 years in the safety profession before returning to construction. During my time in the safety profession, I held the Certified Safety Professional certification (1995-2010), and I maintained a specialty in construction safety and health because of a personal interest. I also taught a college course in construction safety and health for a university for four years. I have taken what I taught during the years in that course and condensed the material into this book.

I hope you enjoy this material and use it on the job daily to prevent or control the hazards found in the construction trade today. It is essential that you control or eliminate risks to ensure you can meet your responsibility to protect workers and return them home to their families and loved ones safe and sound. I begin this book with what you need to know about the Occupational Safety and Health Administration.

For more information on this or my other books please visit my webpage at https://www.fredefanningwriter.com.

Chapter 1-The Occupational Safety and Health Administration

There is a Federal Agency whose focus is on preventing workplace injuries and illnesses. The Occupational "Safety and Health (OSH) Act of 1970 was signed by President Nixon on December 29, 1970, and became effective on April 28, 1971 (Della-Giustina, 2000)." "The Act authorized the Occupational Safety and Health Administration (OSHA) to regulate private employers in the 50 states, the District of Columbia, the Virgin Islands, American Samoa, Puerto Rico, Guam, and the Trust Territories of the Pacific Islands (Della-Giustina, 2000)." The original intent of the OSH Act was for OSHA to approve State programs and pay a percentage of the cost to operate those programs.

The OSH Act covers all employees in the U.S. except for family businesses and the self-employed. Family businesses are exempt if they only employ immediate family members. If a family business employs people outside the immediate family, the business is covered by OSHA. The OSH Act does not require protection for customers or the public.

The OSH Act focuses on the company and goes so far as to make the employer responsible for guaranteeing that an employee will always observe OSHA requirements. If the employee does not comply, the employer can be cited for the employee's failure to follow standards. When it comes to multi-employer work sites, OSHA can cite all employers with employees exposed to the hazards.

Section 5, (a), (1-2) of the OSH Act is referred to as the "General Duty Clause" (The OSH Act, 1970) and it outlines the general duty of each employer:

(a) Each employer -

(1) shall furnish to each of his employees' employment and a place of employment that is free from recognized hazards that are causing or are likely to cause the death or serious physical harm to his employees.
(2) shall comply with occupational safety and health standards promulgated under this act

Section 5, (b) identifies requirements for employees; however, I am not aware of any instance when OSHA has cited an employee for an unsafe act. The act states that "Each employee shall comply with occupational safety and health standards and all rules, regulations, orders issued pursuant to this Act which is applicable to his own actions and conduct." The employee responsibilities generally include:

- Read the OSHA Poster
- Follow safe work practices.
- Cooperate with OSHA inspectors
- Report hazardous conditions to a supervisor.
- Follow the Employer's safety and health rules.
- Wear and use all required gear and equipment.
- Report hazardous conditions to OSHA, if the employer did not fix them.

If a union represents the workforce OSHA will work with them rather than the individual employees. OSHA will also invite a union officer to accompany them on any site visits.

OSHA also provides consultation assistance to those employers that request them. These services are provided:

- at no cost to employers
- with no penalties proposed or citations issued
- for smaller employers with hazardous operations
- by state government agencies or universities employing professional safety and health consultants
- with possible violations of OSHA standards not reported to OSHA enforcement staff unless the employer fails to eliminate

or control any serious hazard or imminent danger

To provide further assistance OSHA maintains an emergency hotline that is 1.800. 321.OSHA (6742). This hotline can be used to:

- report a workplace hazard.
- request an OSHA publication
- request information from OSHA
- file a complaint about a workplace hazard.
- report workplace fatalities or hospitalizations

State Programs

Originally OSHA intended for each state to operate state programs; however, that did not happen. There are 22 states with OSHA-approved state plans. Those states with OSHA-approved programs can establish standards for employers within their state: however, these standards must be at least as effective as the federal standard for a particular area.

Regulations and Standards

OSHA develops standards through the same processes used by most U.S. Federal Agencies. Employers must follow the standards that OSHA develops. Where OSHA does not have standards, companies are responsible for following the OSH Act's General Duty Clause addressed earlier in this chapter. OSHA publishes standards within the Code of Federal Regulations (CFR). The 29 series of the CFR is the Labor Category. There are two types of standards known as horizontal and vertical. Horizontal standards cut across many industries while vertical standards focus on one industry. There are five essential parts to the 29 series. In the construction industry, there are three commonly used parts that include:

- 29 CFR Part 1904 - Recordkeeping
- 29 CFR Part 1910 - General Industry
- 29 CFR Part 1926 - Construction

The 29 CFR Parts 1904 and 1910 are horizontal standards while Part 1926 is a vertical standard specifically for the construction industry.

OSHA Standards may be obtained in public libraries or at the Government Printing Office website or from commercial printers and booksellers. The electronic version can be purchased or accessed at osha.gov.

In addition to the parts of the CFR there are also consensus standards that are defined as:

> "According to 29 USC § 652 (9) [Title 29 Labor; Chapter 15 Occupational Safety and Health] the term "national consensus standard" means any occupational safety and health standard or modification thereof which (1), has been adopted and promulgated by a nationally recognized standards-producing organization under procedures whereby it can be determined by the Secretary that persons interested and affected by the scope or provisions of the standard have reached substantial agreement on its adoption, (2) was formulated in a manner which afforded an opportunity for diverse views to be considered and (3) has been designated as such a standard by the Secretary, after consultation with other appropriate Federal agencies" (Legal Definition, 2013).

There are several consensus standards developed by organizations that are cited by OSHA reference in the standards. These consensus standards include:

- Life Safety Code
- National Electric Code
- National Fire Protection Association Codes
- American National Standards Institute Standards
- American Society of Mechanical Engineers Standards
- American Conference of Governmental Industrial Hygienists Standards

In addition to standards, there are Letters of Interpretation. On its website OSHA explains them as:

"OSHA requirements are set by statute, standards and regulations. Our interpretation letters explain these requirements and how they apply to circumstances, but they cannot create additional employer obligations. These letters constitute OSHA's interpretation of the requirements discussed. Note that our enforcement guidance may be affected by changes to OSHA rules. Also, from time to time we update our guidance in response to new information" (Letters, 2013).

Competent Person

"The term "Competent Person" is used in many OSHA standards and documents. OSHA defines a competent person as "one who is capable of identifying existing and predictable hazards in the surroundings or work conditions which are unsanitary, hazardous, or dangerous to employees, and who has the authorization to take prompt corrective measures to eliminate them"" (Competent, 2014).

The competent person obviously needs to have special training and knowledge with respect to soil classification, the use of protective systems, and the requirements of the standard for excavation, trenching, and shoring. Special training gives them the skills to identify and eliminate hazards. It might help to look at an example of the knowledge a competent person must know. Excavation is a dangerous operation, before any excavation begins the competent person should:

- Identify utilities.
- Determine the soil condition.
- Determine needed safety equipment.
- Ensure there are adequate entrances and exits.
- Manage the construction of protective systems.
- Handle tests for oxygen, hazardous fumes, and toxic gases

Furthermore, if we take the example a little further a competent person must make daily inspections of the excavation and the area around it before work starts. He or she must also conduct an inspection after a rainstorm, high winds, or any other occurrence that may create or increase the hazards that workers may be exposed to. The competent person must have the authority to remove workers from the excavation or shut down the job if he or she finds evidence of a possible cave-in, indications of failure of protective systems, hazardous atmospheres, or other dangerous conditions. From this example, you can see how important the duties of a competent person are.

Accident Reporting and Investigation

The 29 CFR Part 1904 spells out the accident reporting and investigation process. *All employers with 11 or more employees* must maintain records of occupational injuries and illnesses. *All employers* must report to OSHA within eight hours any accident that results in a fatality or in-patient hospitalization of three or more employees. Businesses in the construction industry must comply with 29 CFR 1904. Officially Reportable Accidents are those that result in injuries or illnesses that occur in the workplace and must be reportable to OSHA. OSHA will investigate fatal accidents. Unfortunately, these investigations can result in citations for noncompliance with standards that are identified during the investigation. These citations may result in fines to the company.

In addition to reportable accidents, there are recordable accidents. A recordable accident is a work-related injury or illness that results in death, loss of consciousness, days away from work, restricted work activity or job transfer, or medical treatment beyond first aid. Two definitions of recordable accidents are:

- an employee fatality
- an employee who is transferred to another job because of an injury.

The OSHA Form 300 is a log; OSHA Form 300A is an annual report of recordable incidents, and OSHA Form 301 is an Injury and Illness Incident Report.

The employer must log accidents on the OSHA Form 300, which has separate areas to record injuries and illnesses. Following that, the employer completes the OSHA Form 301 on each accident within seven days of its occurrence. OSHA provides forms 300 and 301. At the end of the year, the information on the OSHA form 300 is transferred to the OSHA Form 300A and then posted the following year. A best practice is to maintain all records for five years.

General Industry Standard

29 CFR, Part 1910 includes standards for General Industry. Even though these standards are not specific to the construction industry, construction businesses are required to comply with them. Because this standard cuts across all industries OSHA does not include these standards part and the standards remain in 29 CFR 1910. The sections that businesses in the construction industry must comply with are in Table 1 below.

Sections of CFR 1910			
Guarding Floor and Wall Openings and Holes	1910.23	Lockout/Tagout	1910.147
Occupational Noise Exposure	1910.95	Powered Industrial Trucks	1910.178
Personal Protective Equipment	1910.132	Bloodborne Pathogens	1910.1030
Respiratory Protection	1910.134	Hazard Communication	1910.1200
Permit-Required Confined Spaces	1910.146		

Table 1 – Sections of 29 CFR 1910 that the construction industry complies with

When OSHA started, it had to develop safety standards. The Construction Safety Standards were initially adopted by the Department of Labor in April 1971 to implement the Contract Work Hours and Safety Standards Act, 40 United States Code (USC) 333. Over the years, these standards led to 29 CFR Part 1926 for the construction industry. Part 1926 is a wide range of activities in construction, alteration, and repairs (Construction Industry, 2013); see Table 2 on the next page.

29 CFR Part 1926			
A	General	N	Cranes and Derricks
B	General Interpretations	O	Motor Vehicles, Mechanized Equipment and Marine Operations
C	General Safety and Health Provisions	P	Excavations
D	Occupational Health and Environmental Controls	Q	Concrete and Masonry Construction `
E	Personal Protective and Life Saving Equipment	R	Steel Erection
F	Fire Protection and Prevention	S	Underground Construction, Caissons, Cofferdams and Compressed Air
G	Signs, Signals, and Barricades	T	Demolition
H	Materials Handling, Storage, Use, and Disposal	U	Blasting and the Use of Explosives
I	Tools, Hand, and Power	V	Power Transmission and Distribution
J	Welding and Cutting	W	Rollover Protective Structures; Overhead Protection
K	Electrical	X	Stairways and Ladders
L	Scaffolds	Y	Commercial Diving Operations
M	Fall Protection	Z	Toxic and Hazardous Substances

Table 2 – 26 Subparts of Part 1926 of 29 Code of Federal Regulation

Violation of Standards

There are several types of OSHA violations. These include:

- Willful - an intentional violation of the OSH Act or plain indifference to its requirements.
- Repeat - meaning two or more substantially similar violations. Can be at different locations, and there is no time limit between the violations.
- Serious – the substantial probability that death or serious physical harm could result.

The most serious violation that OSHA cites is a willful violation. Employers can be required to pay a minimum of $5,000 or a maximum of $70,000 for each violation. If the violation results in the death of an employee, the employer could be fined up to $250,000 or imprisoned for up to six months if criminally convicted.

OSHA may or may not issue citations. If they do, the citation tells the employer and employees the standard allegedly violated and the proposed time for abatement or correction. The employer must post a copy of each citation at or near the place where the violation occurred, for three days or until the violation is corrected, whichever is longer.

Inspections

OSHA enforces the standards through compliance inspections conducted by an OSHA employee called a "Compliance Officer." This person conducts site inspections and writes citations for violated safety standards. OSHA has several inspection priorities that include:

- Imminent Danger Inspections - conducted where any condition exists where there is a reasonable certainty that a danger exists that can be expected to cause death or serious physical harm immediately or before eliminating the risk through normal enforcement procedures.
- Fatalities and Catastrophes Inspections – conducted because of an accident that results in hospitalization of three or more employees.
- Employee Inspections – conducted because of employee

complaints or referrals.

- Programmed High Hazard Inspections – conducted of high hazards identified through analyses.
- Follow-Up Inspections - conducted to follow up on previous inspections.

During inspections, all contractors and employee representatives will, at some time during the examination, be informed, why a focused or comprehensive review is being conducted. OSHA wants to hear from employees or their representatives if they are unionized.

No one in management, except the company's designated representative; should engage in any substantive discussion with an OSHA inspector. Things said in passing can cause a violation. This is like a law enforcement search and employers should know their rights.

Training Requirements

Many standards developed by OSHA explicitly require the employer to train employees. Contractor or company personnel can conduct training. OSHA standards make it the employer's responsibility to limit certain job assignments to certified, competent, or qualified employees. The person who is selected or assigned by the employer or the employer's representative as being qualified to perform specific duties is Competent. There are specific courses that provide training for "Competent Persons".

It is not easy to find all the training requirements within the lengthy and complicated OSHA standards. OSHA's training guidelines follow a model that consists of determining if training is needed, identifying training goals and objectives, developing learning activities, conducting learning activities, conducting training, and improving the training program. OSHA's Voluntary Training Guidelines (OSHA Booklet 2254) are a useful resource. To identify required training a company employee goes through the standards and identifies training required for the work being done.

If the problem is solved, in whole or in part, by training then the next step is to determine:

- What to train?
- Who to train?
- When to train?
- Who conducts training?
- Where to train?

Do not try to provide training to solve a problem that is not caused by a lack of training.

Summary

As the Federal Agency with responsibility for preventing workplace injuries and illnesses, OSHA is an excellent resource for construction sites. OSHA does much more than compliance. Long before an inspection is conducted, they will help if contacted. They conduct assistance visits, provide training and publications, and can give advice and answer questions right over the phone. The best thing about using them is that they are the primary source and are a free or low-cost option. The best way to use OSHA is through their website at http://www.OSHA.org/.

Chapter 2–Personal Protective Equipment (PPE)

One way of protecting workers is to have them wear protective equipment. Wearing protective equipment is often not the best way because it depends on the employee wearing the equipment properly when needed. As you can imagine this does not always work well. OSHA 29 CFR, Part 1926, Subpart E: Personal Protective and Life Saving Equipment is the primary source for these hazards and require employers to implement a PPE program that includes:

- Assessing the workplace for hazards
- Requiring employees to wear selected PPE.
- Informing employees why the PPE is necessary, how and when it must be worn.
- Select appropriate PPE to protect employees only for hazards not eliminated.
- Using engineering and work practice controls to eliminate or reduce hazards before using PPE.
- Training employees to use and care for their PPE, including recognizing deterioration and failure.

Conducting a Job Hazard Analysis provides the information needed to assess workplace hazards. Focus on identifying all the tasks that an individual worker would perform and then identify the risks associated with each task. Follow that by identifying measures to control or eliminate each of the hazards. Always use engineering and work practice controls before choosing PPE. Engineering and work practice controls are much more efficient than using PPE.

There are several ways to use engineering controls first to control or eliminate hazards. Substituting less harmful substances for harmful substances can reduce danger. Isolate the task by using such things as barriers or guards or lockout or tagout of energy sources. Another engineering control is the use of ventilation to remove the hazard.

Modifying a piece of equipment can also be used to control or eliminate hazards. Any one of these techniques is a way to use engineering controls and is more effective in controlling risks because after implementation they do not rely on workers; however, workers must be prevented from removing or disabling them.

If the company cannot implement engineering controls, then the company should use work practice controls. This option includes changing work habits, improving sanitation and hygiene practices or changing job procedures. Changing the actual procedures, the worker performs to complete the job is the intent. Another option is to modify or rotate work schedules. Changing work rotations and schedules is not as active as engineering controls because workers must act.

Another option is to use administrative controls. These involve changing standard operating procedures, implementing workplace rules, and using PPE. It is common practice for employers to provide PPE for the workers, train the workers in how to use the equipment, provide a clean storage location, and provide a means to replace PPE. OSHA could cite the employer if a worker is not using the proper PPE even if the employee provided that PPE. These controls rely more heavily on training than the others. It is important to document the training and examination for each worker as a record. Only when other control methods cannot be used should workers use PPE.

Only use PPE that meets the American National Standards Institute standards for the piece of equipment. Always develop a standard operating procedure or policy letter that outlines the results of the job hazard analysis and make sure everyone understands the hazards and how to control them. There are a variety of PPE to use.

Head Protection

Head Protection provides protection for the workers from falling objects; hitting their head against objects and preventing contact with exposed electrical wiring or charged components. Head protection comes in the form of hardhats or bump caps. Bump caps have limited use to prevent injuries from hitting the head against an object and are not used in lieu of a hardhat. Hardhats must comply with American National Standards Institute (ANSI) Z-89.1-1986 that requires a hard outer shell and a shock-absorbing headband. Workers should not wear hardhats backwards unless the hat was designed that way. Limit the use of stickers and paint on hardhats because they can hide flaws or deformities in the hardhat. Require workers to clean hardhats periodically with soap and water. Limit the soap to a mild brand because some strong cleaning chemical agents can weaken the shell and damage the hardhat's ability to provide electrical resistance. There are several classes of hardhats, and it is crucial to provide the right kind, see table 3. It is important to replace damaged hardhats when damaged.

Class	Description
A	- General service (building construction, shipbuilding, lumbering) - Good impact protection but limited voltage protection
B	- Electrical / Utility work - Protects against falling objects and high-voltage shock and burns
C	- Designed for comfort, offers limited protection - Protects against bumps from fixed objects, but does not protect against falling objects or electrical shock

Table 3-Classes of hardhats

It is important to purchase the correct class of hardhat for the hazards. Finally, workers must be trained on the hazards, how the hardhat protects the worker, how to maintain and clean the hardhat, and how to get it replaced if damaged.

Foot Protection

Foot protection provides protection of the foot or more precisely the toes, heel, sole, or the Metatarsal of the foot. Foot protection is also used to prevent slipping and to control electrical conductivity. Foot protection is also called footwear, shoes, or boots. When any of the following hazards have been identified foot, protection provides protection:

- Hot surfaces
- Wet surfaces
- Static electricity
- Slippery surfaces
- Splashing molten metal
- Sharp objects such as nails or spikes
- Heavy objects that might roll onto or fall on the workers' feet.

Electrically nonconductive foot protection protects against the buildup of static electricity and protects equipment and prevents fires or explosions caused by a spark. This protection is nonconductive and prevents the footwear from completing an electrical circuit. Impact-resistant footwear provides protection for the toes and Metatarsal from objects falling or rolling onto them. Heat-resistant soles on the shoes or boots protect against hot surfaces that a worker might experience when roofing or paving. There is also footwear that provides metal insoles to protect against puncture wounds from stepping on objects that might puncture the sole and injure the foot. A flame-resistant flap covering shoelaces on footwear are used by welders to keep the laces from catching fire. It is essential to provide footwear that is fashionable, stylish, and comfortable. Otherwise, the workers will not wear the footwear. To be effective, the worker must know the hazards and how the footwear provides protection. They must be trained to maintain and clean the footwear and how to replace them if they are damaged.

Eye Protection

Eye protection shields the eyes from impacts from work, splashes of chemicals, dust, and burns from radiant light. Eye protection includes glasses, goggles, and face shields. There are various types of goggles that protect from dust, impact, or chemical splash. OSHA publication 3151 is an excellent resource for this type of protection. Another exceptional resource is ANSI Standard Z87.1-2010. Eye protection protects the eyes from:

- Splashing molten metal
- Corrosive gases, vapors, and liquids

- Potentially infectious materials or hazardous chemicals that splash.
- Dust and other flying particles, such as metal shavings or sawdust that includes dust and particles created from using compressed air to clean equipment or clothes.

To get workers to use eye protection it must be comfortable, not restrict vision or movement, be durable and easy to clean, and not interfere with other PPE. They must also be fashionable. These also are the main reasons workers use for not using eye protection.

Ordinary glasses do not provide protection; however, prescription glasses can be purchased with side shields and protective lenses. These glasses can be expensive, and most companies do not pay the extra expense. Companies often choose to use goggles that fit comfortably over corrective glasses without disturbing the glasses or goggles with prescription lenses mounted behind the protective lenses. Workers can place the prescription lenses in different pairs of goggles. In the following paragraphs, the different types of eye protection are addressed.

Protective glasses are used to prevent objects from striking the eyes and light radiation. These glasses must provide direct and side protection. Workers must always use side protection or shields. Protective glasses do not protect the eyes from dust or liquid spills. ANSI standards describe the protection provided by the eyeglasses.   Experts recommend that worker's wear eye protection out of doors to protect their eyes from Ultra-Violet A (UVA) and Ultra-Violet B (UVB) rays. This radiation is the greatest between 10:00 am and 3:00 pm. Protection usually comes with sunglasses, but not all sunglasses provide protection. The worker's surroundings can change the amount of radiation a worker is exposed to. Snow increases the radiation reflecting from the snow. In the summer, grass leaves can increase exposure. The Food and Drug Administration mandates standards for UV protective sunglasses. ANSI Standard Z80.3-2010 also applies to this protection.

Goggles are preferred eye protection. Goggles can protect from impact, dust, liquid splashes, or light radiation. Unlike protective glasses, goggles also protect the eyes and the area immediately around the eyes from impact, dust, splashes, and light radiation. As noted, earlier goggles can also fit over prescription glasses. Stores sell goggles that provide UVA and UVB ray protection.

Welding Safety Goggles are a particular kind of goggle that protects eyes from intense concentrations of light in the form of UV and infrared (IR) created by welding. These also come in the form of a glass shield in a welder's hood. Welding hoods protect the face and eyes from flying sparks, metal spatter, and slag chips produced during welding, brazing, soldering, and cutting. Impact eye protection should always be worn underneath the welding hood to protect the eyes when the hood is lifted to chip the slag off the weld bead. The protection provided by welding goggles or hoods is related to the type of welding being done. They are not one size fits all.

Often face shields are used as eye protection. It is important to remember that face shields do not meet the ANSI Z87.1-1989 standards for impact protection. Workers should wear impact glasses or goggles underneath the face shield to protect from impact. The face shield protects the face from dusts and splashes or sprays of hazardous liquids. When using a face shield with impact protection underneath thorough protection is achieved. Use face shields when grinding, sanding, or cutting.

Hearing Protection

Hearing Protection protects the inner ear from the damage caused by sounds present in the workplace. Hearing protection equipment includes earmuffs, earplugs, and canal caps. Control noise by engineering controls or reduced exposure to the worker whenever possible. When engineering and workplace controls are not feasible companies can use hearing protection. Provide hearing protection when a worker's noise exposure exceeds an 8-hour time-weighted average sound level of 90 decibels on the A scale. Sound is measured by a sound level meter. The need for hearing protection can be roughly compared to noise in the workplace that requires workers to yell over the noise to be understood. There are sound reduction ratings for hearing protection that have been developed by the Environmental Protection Agency. Find the latest information on these ratings at http://www2.epa.gov/aboutepa/epa-establishes-new-noise-label-program. The ratings show how much the sound level is reduced using the hearing protection. The higher the rating, the more protection provided.

Once the sound level requires hearing protection it is important to determine what type is best suited for the workplace? Hearing protection must be comfortable to wear, not restrict direction or warnings from being heard, be durable and easy to clean, and not interfere with other PPE if workers are expected to use them. This includes the eye glass arm breaking the seal of earmuff when in place. These are the reasons workers give for not using hearing protection. Using dirty ear plugs can cause infection, replacements must be readily available.

Hand Protection

Hand protection provides protection against burns, bruises, abrasions, cuts, punctures, and chemicals. All types of gloves provide hand protection. Gloves come in metal mesh, leather or canvas, fabric, and coated fabric, liquid resistant, and rubber. Common types of gloves include:

- Leather and canvas are the most used gloves on construction sites to protect against dirt, abrasion, and cuts.
- Welding gloves are used to protect against the hazards of welding and include cuffs that go up the forearms.
- Fabric and coated fabric gloves protect from dirt and abrasion.

If hazards confirm hand protection is needed, gloves are chosen to protect against them. To get workers to use gloves they must be comfortable, not restrict movement of hand and fingers, be durable, and easy to clean. The employer must have a system in place to replace gloves immediately when damaged.

Body Protection

Body protection protects the body from the intense heat; splashes of hot metals and other hot liquids; impacts from tools, machinery, and materials; cuts; and hazardous chemicals and materials. This protection is for parts of the body exposed to possible injury. The more common types include cooling vests, aprons, jackets, coveralls, and full body suits. Once the hazards are identified the company determines the type of body protection required. For welders, this clothing includes:

- Apron and chaps made of flame-resistant material.
- Chaps and cape sleeves made of flame-resistant material. The cape sleeve is like the upper part of a shirt that covers the length of both arms, across the shoulders and the upper part of the back and chest.
- Flame resistant gloves called gauntlets. These gloves include a cuff that goes up the forearm a couple of inches.
- A cap or hat made of flame-resistant material.

Special clothing is also required when working with live high voltage. National Fire Protection Standard 70E is a good resource for this hazard. This clothing consists of flame-resistant clothing, full body suits, gloves, and hoods.

Remember to adjust the amount and type of clothing based on the size, nature, and location of the work. To encourage workers to use the equipment they must fit properly, can be cleaned, and do not restrict body movement.

Respiratory Protection

It is best to bring in an Industrial Hygienist Consultant to determine respiratory hazards. Specific testing must be performed to determine the risk and the corresponding respirator to provide the protection. Respiratory hazards are inhaled fumes, mist, dusts, and fibers. Respiratory protection stops the hazard from entering the lungs.

There are two types of risks: short term and long term. Short term risks make the worker ill right away. Long term risks may take years after the exposure to make the worker ill.

If respiratory protection is required, the company must implement a respiratory protection program. The employee must pass a physical prior to being authorized to wear a respiratory. Respirators are extremely complicated and include a variety of filters specific to the risk. The respirator is fit tested specifically to the worker. Workers must be trained in the proper maintenance and cleaning.

In addition to hazards, workers may be exposed to situations where oxygen is deficient. In these cases, oxygen will have to be provided with the respirator. It is easy to see why a called is placed to an Industrial Hygienist Consultant for this hazard.

Summary

PPE provides protection from the hazards identified in a Job Hazard Analysis. Purchase only PPE that meets ANSI standards for a particular hazard. Medical services are necessary for some PPE that might involve placing an employee on a medical surveillance program. It is also crucial that workers be trained on the hazards, how the PPE protects them, how to maintain and clean the PPE and how to get PPE replaced if damaged. Supervisors must make sure workers wear required PPE. The 2002 version of the OSHA Publication 3071 titled "Job Hazard Analysis" is an excellent source for additional information on conducting this type of analysis and can be found at http://www.fs.fed.us/r8/allhazardresponse/Safety/documents/JHA-OSHAJobHazardAnalysis.pdf. OSHA Publication 3155 titled "Personal Protective Equipment" is another excellent source of information and can be found at http://www.osha.gov/Publications/osha3151.pdf.

Chapter 3–Signs, Signals, Barricades/Motor Vehicles and Mechanized Equipment/Rollover Protective Structures

Moving equipment creates hazards in and around the construction site for equipment operators and non-operators alike. Whenever workers encounter equipment, the worker is usually injured. That is why it is essential to consider workers when routing and moving equipment on site.

Part I Signs, Signals, and Barricades

Workers perform construction work in and around people and traffic. In these situations, it is important to highlight risks and make them prominent. There are several ways to make the dangers known, but OSHA describes three that include signs, signals, and barricades. OSHA 29 Code of Federal Regulation, Part 1926, Subpart G: Signs, Signals and Barricades is the primary source for these hazards (Signs, 2014). Furthermore, ANSI Z35.1-1968, Specifications for Accident Prevention Signs contains additional rules. Familiarity with both standards is important.

Signs warn of hazards. They can be posted temporarily or permanently at the location of the hazards. There are danger and caution signs. Danger signs are for immediate hazards and use red as the dominate color with black outline on borders and white lower panel for additional sign wording. Caution signs warn of potential hazards or unsafe practices. Caution signs use yellow as the dominate color, with black upper panel and borders, yellow lettering of "caution" on the black panel, yellow lower panel with black lettering for additional wording. Traffic signs must also meet ANSI D6.1-1971, Manual on Uniform Traffic Control Devices for Streets and Highways (Accident Prevention, 2014).

Signals are moving signs. Workers and devices provide signals that warn of existing hazards. Road and highway construction sites use signals in the form of flagmen or flashing lights. When signs cannot provide the necessary protection signals can be used. Flag workers use flags or warning paddles. The paddles they use must be 18 inches square and have a red light for night usage. Flag workers wear red-, orange-, or lime-colored garments during the day and garments with reflective material at night because they are exposed to traffic. Since flag workers are exposed to motor vehicle traffic, they must follow ANSI D6.1-1971, Manual on Uniform Traffic Control Devices for Streets and Highways (Accident Prevention, 2014).

Barricades are obstructions that deter workers on foot or in vehicles from passing through to a hazardous area. Barricades can be plastic barrels, tall cones, sawhorse type, or actual bollards in the ground. Barricades must be in accordance with ANSI Standard D6.1-1971, Manual on Uniform Traffic Control Devices for Streets and Highways.

Part II Motor Vehicles and Mechanized Equipment

Workers that operate conventional vehicles like pick-up trucks, vans, and cars usually with a standard driver's license. All other motorized vehicles and mechanized equipment require special training and licensing to operate. OSHA 29 Code of Federal Regulation, Part 1926, Subpart O–Motor Vehicles, Mechanized Equipment, and Marine Operations is the primary source for these hazards (Motor Vehicles, 2014). The state law is another primary source for operating motor vehicle and mechanized equipment on public roads and streets.

It would be nice to hire the drivers and operators that are needed already trained, but that usually does not always happen. There are specific training requirements for material handling equipment, more commonly known as forklifts. Most pieces of equipment need a company representative to give the drivers or operators a test to ensure they can operate the vehicle or equipment properly. A good training example is a five-week school to operate a bulldozer. During that training, the operator is also trained on the dump truck and semi-tractor/trailer that hauls the dozer. At the end of those five weeks the worker is still learning, and they must relearn each time they do something new or run a different size dozer.

It is imperative that the drivers and operators first know how to operate a vehicle or piece of equipment properly. Some basic requirements that drivers and operators must be familiar with are:

- Batteries
- Refueling
- Tire cages
- Jump starting.
- Using starter fluid
- Blocking and cribbing
- Parking brake and wheel chocking
- Marking of unattended equipment at night

Most vehicles and equipment used in construction are large and heavy. They use diesel engines and need battery power, usually 24 volts, to start.

Operators maintain batteries and check them daily wearing leather gloves. Operators should not wear jewelry when working on batteries. They wear a face shield when adding water to batteries. In accidents, the battery box and the batteries inside can be damaged and may need attention. Equipment usually comes with a slave cable that is a battery connection so that jumper type cables are usually not used. This cable is plugged from one piece of heavy equipment to another to provide electricity for starting.

In extreme cold weather, operators may use starting fluid to help the engine start. This fluid is ether and can be highly flammable. Workers should not use starting fluid around open flames or while smoking. Using too much starting fluid can damage the engine, operators should use only enough fluid to get the engine started.

Refueling of vehicles and equipment is done at the end of the day. The fuel truck usually comes to the equipment and requires the lifting of a heavy hose into the tank opening. Operators are the one that often climb to reach the fuel tank to refuel. It is always best to bond the vehicle or equipment to the fuel truck with a cable and then ground the fuel truck to the earth to prevent static electricity. Do not allow any smoking around refueling operations. Fuel truck operators should carry spill kits with them to prevent fuel from leaking into the ground or a nearby storm drain.

Many vehicles and pieces of equipment use large tires that are not on a rim like the tires on a personal car, but rather on a split rim. Some of these tires can be massive, and another piece of equipment might be needed to remove them to prevent workers from being injured. Big pry bars are used to remove and install tires on the rim. Workers can experience injuries when working with these large bars so they must use care. Once the tire has been repaired and, on the rim, air can be placed into the tire; however, the split rim can come apart with pieces blown away at speeds that can kill a worker. The worker should place the split rim wheel in a tire cage before inflating. The tire cage is fastened to a building or large object to hold it from moving. Then if the tire explodes the cage will take the brunt of the damage saving the worker.

Most of the equipment on construction sites have removable cutting edges. The edges take the damage while protecting the actual piece of equipment. Once worn, replace the cutting edge before damaging the piece of equipment it is bolted to. Changing cutting edges is an extremely dangerous undertaking and is treated that way. The first danger comes from the cutting edges being bolted to massive pieces of equipment that are made to raise and lower. Before any work is done on the cutting edges, it is essential that the equipment is raised to an appropriate level for the work and then blocked or cribbed with something strong enough to hold the weight. An example is a dozer blade. The blade is raised to a level so a worker can remove and replace the cutting edges. The blade is blocked at that height so it cannot fall on the worker. The workers also place blocks under vehicle wheels or tracks to keep the vehicle from moving. The cutting edges themselves can also be heavy and many of the bolts rust on so care must be taken here, as well. Make sure the tools are large enough to the job.

Vehicles and equipment often need to be parked on or near construction sites. When parked or not moving the operator should set the parking brake and block the wheels or tracks. Except in extreme cold weather when setting the parking brake, could result in it freezing in the locked position. Also, if the equipment is to be parked overnight it needs to be marked with reflected surfaces or lights so it can be seen by others in the dark.

Moving large vehicles and equipment can be dangerous especially when in reverse. While backing, signal alarms are used to let others in the general vicinity know when the vehicle is backing up. Workers on the ground around equipment were injured and killed by being backed over before these alarms were mandatory. In addition to the back-up alarms, ground guides should be used when the driver or operator has an obstructed rear view. *EHS Works* reported that from "2003 to 2010, 443 workers were killed on construction sites by a vehicle or mobile equipment. Of those cases, 143 involved a vehicle or mobile equipment that was backing up" (EHS, 2014).

Vehicles and equipment with cabs should have all glass free of distortions or cracks. Windows require some effort to keep clean, but it can be done.

For vehicles and equipment that are tall or with raised attachments take extra care near power lines. Signs should be placed on the ground under the overhead wires to warn operators.

Motor vehicles and equipment that are used only on job sites still need to have working emergency brakes, parking brakes, headlights, and taillights. Properly maintain equipment even though not used on public roads or streets. Vehicles and equipment provided with a windshield must keep it in good working order. If it also has powered wipers, the wipers must be kept in good working order too. It is also important for drivers and operators to wear seatbelts. Even in something as slow as a bulldozer belts can prevent injury and death. To keep things in good working order requires drivers and operators to do pre-operational checks to identify anything missing and not working. This information is then passed on to the supervisor.

Companies use many types and sizes of tractors on construction job sites. Most of these are agricultural tractors with an engine that has 20 horsepower. These tractors are used for pulling, propelling, or driving implements. There are also industrial tractors that have an engine with more than 20 horsepower used for landscaping, services, loading, digging, and mowing.

In the construction industry, other vehicles known as Material Handling Equipment or MHE are used. These include earthmoving equipment like dozers, graders, and scrapers. The employer is also required to re-train operators if they identify issues with their abilities. One last note on this equipment is that it is often used to clear sites that involve hazards from irritant and toxic plants. Workers should be provided with cabs to protect them from these hazards.

Most people are probably familiar with MHE when it refers to forklifts. Forklifts are also called powered industrial trucks. There are seven classes of forklifts, see table 4. Each has unique characteristics and inherent hazards.

Classes of Forklifts	
Class I	Electric motor rider truck.
Class II	Electric motor narrow aisle truck.
Class III	Electric motor hand trucks or hand/rider truck.
Class IV	Internal combustion engine truck with solid or cushioned tires.
Class V	Internal combustion engine truck with pneumatic tires.
Class VI	Electric and internal combustion engine truck tractor.
Class VII	Rough terrain forklift.

Table 4-Classes of Forklifts

OSHA has specific training standards for forklift operators that require both classroom and hands-on instruction followed by an examination. OSHA 29 CFR 1910.178 is the standard that requires employers to train workers (Powered Industrial Truck, 2014).

Part III Rollover Protective Structures

All MHE require rollover and overhead protection. Without a Rollover Protective Structure (ROPS) the operator might be crushed if the machine rolls over. The ROPS should be factory installed. Manufacturers make ROPS with a grid or mesh type material between solid materials. OSHA 29 Code of Federal Regulation, Part 1926, Subpart W – Rollover Protective Structures; Overhead Protection is the primary source for these hazards (ROPS, 2014).

If the operator will use the equipment for site clearing, the ROPS must include overhead and rear canopy guards made of 1/8 steel plate or ¼ woven wire mesh with openings less than 1 inch. Remember two things:

- Do not remove a ROPS from a piece of equipment.
- Do not build and install a ROPS. The ROPS must have the manufacturers or fabricators name and address on the structure.

Summary

This chapter contains a wide variety of information, but each piece is essential to effective construction safety program. Signs, Signals, Barricades highlight hazards, so the danger is known. Motor Vehicles and Mechanized Equipment creates an entire set of dangers when they are mixed with workers. Rollover Protective Structures are a means to control the danger that equipment operators face if equipment rolls over. The basics remain the same: make sure workers are trained to do the job, aware of the hazards, and empowered to prevent accidents and injuries.

Chapter 4—Hand and Power Tools

Construction workers have little concern with the hazards associated with hand and power tools. This level of concern is probably because most workers have these tools at home and think they know how to use them. This attitude means accidents are likely. The fact is that many construction workers are injured by hand and power tools every year. OSHA 29 Code of Federal Regulation, Part 1926, Subpart I: Hand and Power Tools is the primary source for these hazards.

It is important to know the risks before using hand and power tools. The cutting edges of these tools can create abrasive objects that fly causing impact and striking injuries. One injury can be impact to the eye. Additional risks include dusts, fumes, mists, vapors, and gases. These hazards can damage both the eyes and lungs. Because most of these tools are powered by electricity the hazards include frayed or damaged electrical cords, hazardous connections, and improper grounding. These risks can lead to electric shock or electrocution. In addition to worker injuries, these tools can lead to damage to facilities and equipment. One example is that fumes, mists, vapors, and gases can create fire and explosive hazards that can damage or destroy.

The primary means of reducing or eliminating hazards is to purchase the right tool for the job and always using the tool only for that purpose. Keeping tools in proper working order and replaced when damaged is also important. Device damage can be reduced by keeping tools clean and stored indoors. Prior to using power tools, workers should make sure guards for bits and blades are in place and working. Workers must also acquire and use the proper PPE.

Hand Tools

Hand tools can cause several injuries. The primary hazards associated with hand tools are the result of not being properly maintained or being misused. Workers should check all the tools in the box and look for visible damage, starting start with chisels and punches. Next is to use them properly.

A hammer strikes chisels and punches on the blunt end or head. This striking causes the head to mushroom or flatten. When striking the mushroom pieces of the mushroom can fly off hitting the user. These tools should have the head ground so that the edge is at a 30-degree angle. Periodic grinding of this edge can prevent mushrooming from creating a hazard. The cutting edge of the chisel or punch must also be ground to keep the edge or point sharp so it will have to be struck less. Workers must use eye protection when using these tools. Eye protection will protect against the flying metal chips. Wearing eye protection is crucial when sharpening the tool too. If using a face shield when grinding, wear impact eye protection under the face shield.

It is important to check the heads and handles of hammers too. Breaking the hammer handle is done by striking the handle on the object being hammered. The impact leads to the handle cracking and may cause the head of the hammer to fly off or splinters from the handle become flying wood. Never tape broken handle. This only makes the situation worse because now the user might not know it is damaged. Always replace broken handles. It is often cheaper to replace the whole hammer rather than just a handle. Damage can happen to the head of the hammer. The head can mushroom, or the claw on a carpentry hammer can crack or break. Mushrooming of the head can be fixed by grinding the head at a 30-degree angle eliminating mushrooming. If the claw-end of the hammer if broken, the hammer should be replaced. Also replace hammers with damaged plastic or rubber heads.

Check screwdrivers for cracked handles, bent shafts, or blunt points. Workers often use screwdrivers as pry bars. They are not built to use in this manner and often bend or break as a result. If the shaft breaks, the worker can injure their hand as the screwdriver slips. Workers trying to straighten the shaft may damage it more. It is better to replace the screwdriver with a bent shaft. The handle and end of the tool can be damaged when using screwdrivers as chisels. The screwdriver point is not made to cut as a chisel edge is. When the end is damaged the tool slips off the screw injuring the hand of the user. A damaged flat tip screwdriver can be corrected by grinding the point to a flat edge again; however, if grinding the point will not work replace the device. Damaged Phillips screwdrivers should be replaced. When an operator uses the screwdriver as a chisel, the handle can break or split. Later when using the screwdriver, the broken plastic can pinch or cut the hand. Remember to wear eye protection when grinding the points of screwdrivers. If wearing a face shield, do not forget to wear impact eye protection under the face shield.

Another item to check are the wrenches. Workers use wrenches on bolts or nuts that are too tight or rusted. In either case, if the user pushes too hard the wrench jaws can bend causing the wrench to slip off the nut or bolt. An injury can occur to the hand or arm as it moves in the direction being pushed. Some workers use a piece of steel pipe pushed over the unused end of the wrench to give extra leverage. This pipe is called a "cheater bar" and can damage and even break the wrench resulting also in injury to the worker. Always replace damaged wrenches.

Shovels, posthole diggers, axes, picks, and rakes usually have wooden handles. These tools can be misused cracking or breaking the handle. When the handle is cracked or broken, the user can fall to the ground or injure the hand. Always replace cracked or broken wooden handles.

Pry bars and crowbars are made to pry objects up or apart; however, it is easy to push or pry something too hard or too heavy causing the bar to bend or damage the edge. This can cause the worker to fall forward injuring the hand or arm. The damaged edge of a pry or crowbar can be ground to the proper edge and used again. Never straighten a bent bar, always replace.

It is necessary to keep the edges of cutting tools sharp. Sharpening drill bits and saw blades can be cheaper than replacing. There is a mount that fastens to a bench grinder that allows the user to sharpen drill bits to the proper angle. Keeping the drill bit sharp prevents it from slipping off the object being drilled. Operators can also sharpen saw blades. Workers can use a file to sharpen the blades of hand saws preventing them from binding. Cutting blades can also be taken to a service or business that sharpens them.

Hand tools are designed for a specific purpose and should only be used that way. Tools should be kept clean and in good working order. Tools that are damaged or broken should be repaired or replaced, depending on the type of tool. The proper PPE should always be used to prevent injuries. Always keep tools:

- Clean and free from rust and dirt
- Free from excess oil or grease
- Free from damage
- Sharp

Jacks, Blocking and Bracing

Jacks can also create hazards on job sites. Before raising a load, users should ensure the jack is rated for the weight of the load. The manufacturers mark the rated capacity on all jacks and that must never be exceeded. Inspecting the jack can make sure it is in good working order and not leaking hydraulic fluid. Issues can be prevented by lubricating and inspecting jacks regularly.

Set the jack base in the center of the load being lifted on a firm and level surface. If a firm level surface is not present, a block can be placed under the base of the jack to level it. Before raising a load make sure, the jack head is placed against a flat level surface on the load, so it will not slip off. If the load does not have a flat level surface, a block can be placed between the jack cap and the load. As the jack is used to lift the weight makes sure, force is applied evenly. Never exceed the jack stop indicator that warns the user the shaft of the jack is at its maximum extension.

When using the jack, immediately block the load right after the lift. Blocks can be wood or metal. Wood is a common material on construction sites and makes a good block. It is best to use several pieces of wood to create the block rather than a single piece of wood.

Keeping a few pieces of wedged wood around can be used to help make sure the block is level.

Use jack stands for the weight of the load being blocked. Marking the weight capacity on the stand helps with using the proper stand. It is also important to inspect the stand for damage before each use.

Power Tools

Power tools are designed to provide action through a source other than human. The power for tools comes from a variety of sources that include electric, pneumatic, hydraulic, liquid-fuel, and powder-actuated. Power tools save time and energy; however, often at a cost. Take, for example, a hammer that strikes a worker's thumb as it is swung to hit the nail. Now compare that danger with the possibility of striking the thumb with the discharging nail from a powder actuated nail gun. The difference is significant. What would be a bruise or perhaps a lost thumbnail from the hammer can be severe damage to the thumb by the nail going into or through it. All power tools increase the potential for injury if an accident occurs.

Power tools include handheld tools and accessories. These tools come in the form of drills, impact drills, circular saws, routers, jigs and saber saws, sanders, nail guns, and staplers. Accessories include lights, heaters, and fans. These tools are dangerous and can often be extremely dangerous.

Using power tools only for their designed purpose can reduce or prevent hazards. Prior to any use, make sure the power tool operates correctly. Replacing damaged or non-working tools is a great idea. Having the proper PPE for the tool can prevent injuries. Workers using these tools should not wear loose fitting clothing and jewelry that can get caught in the machines moving parts. When the power tool is not being used eliminate the hazards by disconnecting the machine from the power source. Also remove power from tools before servicing, cleaning, or when changing accessories. Keep fingers off the switch button when carrying powered tools. Keeping tool bits and blades sharp and clean reduces binding and jumping. Lastly secure the work with clamps or a vise freeing both hands to operate the tool.

To reduce the hazards of power tools those equipped with a constant pressure switch that shuts off power upon release are preferred. The switch requires the user to maintain one hand on the tool while it is operating. Unfortunately, this switch is not present on all power tools. Circular saws, grinders, chain saws, and drills will have this switch. Tools like routers, planers, laminate trimmers, shears, and jig saws usually will not have this type of switch and the user is left with the on-off switch. More care must be taken with tools without a constant pressure switch that more dangerous than those with the switch.

Many of the tools in building use electricity as a power source. These tools must be double insulated. Read the manufacturers instruction to determine if a tool is double insulated. The user will provide power to the machine through an electrical extension cord.

The worker is in danger of contacting electricity through contact with a damaged extension cord. Extension cords used on construction sites require a Ground Fault Circuit Interrupters (GFCI). The GFCI prevents the most frequently occurring fault. More can be learned about the GFCI in Chapter 6s.

Do not use the cord to carry, hoist, or lower tools. This method can damage the cord. Yanking the cord from the outlet rather than walking up to the outlet and disconnecting the cord can also cause damage the cord. Yanking the cord or carrying or hoisting tools with the cord can result in the wires pulling out of the plug end too.

Extension cords often encounter water, fuel, heat, oil, or sharp edges. All of these can damage the cord. On some job sites these cords are raised above the floor or ground level to keep them away from water, fuel, oil and prevent workers from walking on them. To do this, the workers use wire trees or wooden boards with cross boards at the bottom that allow it to stand and a cross board at the top to carry the cords. Construction sites are made safer by this method also removing tripping hazards. Before using, inspect cords, only use cords without damage, and those that have the third wire ground prong in place.

Many power tools use abrasive wheels that create additional hazards. The primary hazard is the wheel encountering the hand or arm. Prevent contact with the wheel by using machine mounts and guards.

Additionally, wheels can explode even when used correctly. Inspect abrasive wheels using the sound test before mounting. Before using a wheel inspect it to identify damage and conduct a ring test to ensure that the wheel is free of cracks and defects. The ring test is remarkably simple. Run a string through the center hole of the wheel and allow the wheel to hang freely from the string. Tap the wheel gently with a light non-metallic instrument. The tap should produce a nice clear ringing sound. When the wheel rings it is safe to mount. If the wheel does not ring, dispose of the wheel. Always use the proper wheel.

Grinding on the circumference or outer edge of the wheel like a bench grinder requires a wheel for designed that purpose. Using the wheel to grind on the flat surface or side of the wheel as with a handheld grinder requires a different type of wheel. Use abrasive wheels for their designed purpose. Grinding on the side of a wheel meant for edge grinding can cause the wheel to break and explode.

Abrasive wheels can also collect metal in crevices of the wheel reducing the cooling that should occur. Correct the clogging by dressing the wheel frequently. A wheel dressing tool has small wheels that grind the surface of the edge removing the metal in the crevices.

All machines are purchased with guards that are designed to protect the operator from the point of operation that includes in-running, nip points, rotating parts, flying chips, and sparks. These guards must remain on the tool. Replacement guards can be purchased to replace broken or missing guards. Dispose of tools that guards cannot be found for. It is best to use PPE in addition to guards and not instead of them.

Users might be hit by a tool attachment or a fastener when using pneumatic tools that are powered by compressed air. Check the manufacturer's information and adjust the air pressure to the recommend pounds per square inch or psi before using. When using pneumatic tools, the hose connection must be positively secured to the device to prevent the machine from becoming disconnected with force. Using safety clips will prevent hoses and attachments from accidentally coming free. Compressed air can injure workers primarily by putting particles in the eyes. Never use compressed air to clean clothing or blow off equipment. Hazards can also be reduced by using a chip guard and PPE.

Fuel gases are the primary hazard for machines that use liquid fuel. Reduce the hazard by using only approved flammable liquid containers. Before refilling the tool, let it cool to prevent fuel vapors from igniting. Always use caution refueling while refueling the tool and clean up all fuel spills.

When using powder actuated nail and staple guns workers should be familiar with the tool and have read the manufacturer's instructions for use. The power for these tools comes from an actual round of ammunition. Machine guards must remain installed to prevent nails or staples from being shot at anything other than the material being fastened. Users should wear the PPE. Keep the rounds away from heat and flames and prevent them from being damaged.

Summary

It is important to purchase the proper tools for the job and to use them the way the manufacturer recommends. Workers should know the hazards encountered with electricity, hand and power tools, and jacks and blocks. OSHA has a web page that addresses these hazards that can be found at https://www.osha.gov/SLTC/handpowertools/. OSHA also has booklet number 3080 titled "Hand and Power Tools" and a second booklet number 3459 titled "Nail Gun Safety: A Guide for Construction Contractors" that provide concise information. Both booklets can be found on the OSHA website at https://www.osha.gov/pls/publications/publication.athruz?pType=Ind ustry&pID=50.

Chapter 5–Welding and Cutting

Welding and cutting are frequently used on construction sites. They bring risks that can be reduced or eliminated with proper use of the equipment and PPE. The basic types of welding include gas and arc. The basic types of cutting are oxygen and arc. OSHA 29 CFR Part 1926, Subpart J: Welding and Cutting is the primary source for these hazards.

Gas welding is slower and easier to control than an electric arc. It uses a gas flame with a rod held over the metal until a molten puddle forms, and the rod melts into the joint. The most modern fuels used with gas welding are oxygen and acetylene; however, Mapp gas and hydrogen are also used. Mapp gas is used with oxygen rather than acetylene.

Arc welding is two metals joined together by generating an electric arc between a covered metal electrode and the base metal. The two pieces of metal are made molten along with the rod creating the weld.

Oxygen and Arc cutting are the most often used. These involve severing or removing metal by a flame or arc. Oxygen and acetylene cutting requires metal be heated by gas flame with an oxygen jet doing the cutting. Arc cutting requires intense heat of an electric arc melting away the metal.

Hazards

In general, there are several hazards associated with welding, which include:

- Heat
- Fumes
- Smoke
- Impact
- Harmful Dust
- Light Radiation

Protective Equipment

It is important to protect body parts in proximity to the welding or cutting and the eyes at a distance from arc welding. Caution must be used to prevent breathing in the fumes from welding or cutting. Preparations should also be made to prevent a fire or to fight a fire. Both cutting and welding involve a lot of heat and things near the work can catch fire.

Welding and cutting can cause sparks and spatter (molten metal) to pop and land on the worker or someone helping them. Welders and helpers can also experience injuries from ultraviolet and infrared ray flash burns. These injuries can be severe. Wearing sturdy and flame-resistant clothing is essential. If flame resistant clothing is not available cotton is an excellent material to wear because it will not melt onto the body like some other materials. An apron and chaps made of flame-resistant material that can prevent sparks and molten metal from landing on the welder's clothing or skin are a good choice to acquire. Chaps and a cape sleeves made of flame-resistant material are also a good choice. The cape sleeve is like the upper part of a shirt that covers the length of both arms, across the shoulders and the upper part of the back and chest. Flame resistant gloves called gauntlets are better than regular gloves. These gloves include a cuff that goes up the forearm a couple of inches. A cap or hat made of flame-resistant material protects the head from sparks and spatter. Remember to adjust the amount and type of clothing to the danger, nature, and location of the work.

The eye and face are at greater risk from the hazards of welding, but proper eye and face protection can reduce the risk. This type of protection varies with the welding being doing. It is critical to protect both the eyes and face with a good welding hood. Welding hoods with glass filter plates protect the worker from arc rays while the hood protects the eyes and face from welding sparks and spatters. The welder raises the hood and strikes the slag covering over the melted metal to check their work. To prevent the slag from popping back and up into the eyes, welders should always wear impact eye protection under the welding hood.

If using oxygen and acetylene, the workers need a good pair of goggles. These goggles will protect the eyes from the heat and brightness of the flame and impact. Goggles do not provide protection for the face; however, this method present minor hazards for the face.

Workers weld and cut in the general vicinity of other work. A shield must be set up between the arc of welding and cutting and workers nearby to prevent burning their eyes with the radiation. These burns are usually not severe but can leave a person with dry eyes that feel like they have sand in them. The symptoms last a couple of days.

Ventilation

No matter the type of welding or cutting being done welders and other working nearby may be exposed to noxious fumes from the melting metals as well as the rods used to melt onto the metal. Arc welding rods should come with a Safety Data Sheet that highlights the dangers. Chapter 13 provides more information on these sheets. It is necessary to provide ventilation for the welder. Either natural or mechanical ventilation can work. On most construction sites, welders work outdoors and can use natural ventilation to move fumes away from them as they work. If welding or cutting indoors employers must provide mechanical ventilation. Never use natural ventilation in a confined space. To ensure adequate ventilation is provided the following factors are relevant:

- Size and configuration of the workspace
- The amount of natural air flow in the workspace
- Number and type of work being done creating the hazards.
- Location of workers' breathing zone from welding or cutting.

If working outdoors or in unfinished facility natural ventilation can be used; however, the amount of natural ventilation must be adequate for the work being done. This can be measured by hiring an Industrial Hygienist Consultant. The general rule is to use natural ventilation when:

- There is more than 10,000 square feet of area for each welder.
- There is a ceiling height of more than 16 feet.

- The space does not have partitions, balconies, or barriers that obstruct the movement of natural ventilation.

Other options for using mechanical ventilation consist of using low and high vacuum systems. The low vacuum system moves large amounts of air at low velocity. Low vacuum systems consist of a hood positioned at a distance from the work area that draws the fumes out of the work area. High vacuum systems consist of extractors located as close to the work as possible to capture and extract the fumes. Both systems usually have a fan that draws the fumes through a filter and then recirculates the clean air back into the work area safe to breathe.

The ventilation in place for the workers will also protect others in the general vicinity.

Fire Prevention

It is also crucial to prevent a fire and be prepared to fight a fire if one occurs. If the facility that has a working sprinkler system, that system can be used as protection from fire; however, if a facility is being built the sprinkler systems will not be operational. Fire extinguishers should also be readily available in either case. A multi-purpose fire extinguisher is best; however, make sure to check the Safety Data Sheet before selecting the extinguisher. If electricity is not present a water fire extinguisher can be effective.

To prevent fires, clear a 35-foot area around welding and cutting of all flammable and combustible materials. It is important to consider flammable gases and vapors not just flammable liquids. Also, when cutting or welding on pipes or other metals items remember that the metal can conduct heat away from the work that might ignite combustible material nearby.

Require the use of Hot Work Permits or Work Control Permits when welding or cutting. The hot work permit or work control permit is a form of job hazard analysis or checklist to ensure workers take the proper precautions to prevent the causes of a fire. These permits also address the possible need to perform an outage of the fire alarm or sprinkler system, if one exists, to prevent it from being activated by the welding or cutting. When a fire alarm or sprinkler system is not present or has been taken out of service a fire watch is necessary. A fire watch is a worker placed near the work who periodically checks the work area to ensure a safe condition is maintained by keeping vigil for stray sparks, ignitions, or other fire hazards. Hot Work Permit will not work when:

- Flammable gases or vapors are present.
- An appropriate fire extinguisher is not readily available.
- Combustible or flammable materials within 35 feet of the work
- Cutting or welding on pipes or other metals conducts enough heat to ignite nearby combustible materials.

It is often necessary to weld or cut near a wall or floor made of wood. In these cases, a heat shield can be placed between the work and the combustible work to prevent the wood from catching fire.

Summary

Welding and cutting are necessary work on most construction sites. When this work is performed with hazards present, they must be controlled. Knowing the hazards lets workers identify specific control measures. It is important to manage risks from injury, inhalation, and fire. The best way to manage the risks is to complete a work control permit, which is a form of checklist. Follow that list and double check precautions. Taking these steps can make cutting and welding safe. A page with specific information can be found at OSHA's web page at https://www.osha.gov/doc/outreachtraining/htmlfiles/welding.html.

Chapter 6–Electricity

Electricity is a necessary part of any construction site to power lights and tools. On most sites, installing electrical wiring and equipment is part of the construction being done. The need for electricity requires the use of temporary wiring and installations that can be hazardous. The dangers increase when using power in extreme weather like rain and snow. OSHA 29 Code of Federal Regulation, Part 1926, Subpart K: Electrical is the primary source for these hazards (29, 2012). The National Electric Code-NFPA 70 published by the National Fire Protection Association is also an important source (National, 2013).

Basic Information

Electricity is energy created by charged particles distributed through wires. It can be direct or alternating current. Direct current is provided through batteries while alternating current comes from the utility and distributed through wires. Low voltage power also comes in the form of static electricity. All three are found on construction sites. A few electrical terms that are handy include:

- Watts - power consumed.
- Amps - volume of electrical flow
- Bonding – joining of two objects.
- Volts - measure of electrical force
- Insulator - substance with high resistance
- Grounding - conductive connection to the earth

Hazards

Electricity is hazardous when a worker becomes part of the flow of energy and serves as the ground. It only takes ten milliamperes to kill a worker. There are four types of injuries:

- Electrical shock and burns.

- Electrocution or death due to electrical shock
- Falls caused when a worker jerks after receiving a shock.

Electric shock occurs when a worker encounters electricity. This can be a minor jolt that many have felt or enough to kill a worker. Often the worker in contact with a tool, wire, or another object with electricity running through it cannot release the object. The operator appears frozen. Electrical current running through the worker's body can result in respiratory or cardiac paralysis. The most dangerous situation is when the charge of electricity passes through the heart. If the worker stops breathing or their heart stops beating the worker can die. Burns are the most common electric shock related injury. Furthermore, some workers that experience a shock fall from a ladder or elevation that they were working at when shocked. This fall can result in serious injuries.

Utility electrical wiring is often overhead around construction sites and creates an overhead danger. Equipment, such as dump trucks and cranes, can contact overhead lines bringing electricity from the power lines to the piece of equipment. Contact with power lines can result in injury or death of the operator or workers who may be on our touching the machine. The machine may also catch fire.

Backhoes, excavators, and drills digging into the ground can contact the underground power lines. Breaking or damaging power lines can result in injury or death of the operator or workers on our touching the equipment. The equipment can also catch fire.

Workers commonly use temporary wiring and extension cords on building sites that causes many hazards. It is important to use wires and cords of the right wire gauge or size. For example, machines plugged into extension cords with wires smaller than the machine can draw more current than the cord can handle causing the cord to catch fire. Overheating can trip the circuit breaker or cause a fire. Another hazard that often occurs is for too many devices plugged into the cord. Overloading causes wires to overheat and catching fire.

It is best to use 3-wire cords designed for hard or extra hard duty. Workers should never use typical two wire household extension cords for construction. Three wire hard or extra hard duty extension cords withstand the rough working conditions present on construction sites. It is critical that no damage exists to the ends of cords and that all three prongs are in place. To work in older outlets workers often cut the grounding prong off. This should never be allowed. The plug at the end of the cord should not be pulling loose from the cord or have wires showing. These cords so commonly used that workers do not recognize the hazard.

Controlling Hazards

There are actions that supervisors and workers can take to control or prevent electrical hazards. Supervisors must conduct a hazard analysis of the construction site that includes identifying the overhead and underground power lines. The supervisor must then make sure these locations are marked. The supervisor must also ensure that workers are trained to prevent electrical hazards. Only electricians should repair and maintain power tools and cords.

It is important for all electrical equipment to be free from recognized hazards, frayed cords, and improperly spliced components. Workers should inspect tools prior to use and not use damaged ones. The arcing parts of tools should be enclosed to prevent sparks, arcs, flames, or molten metal. A spark from these parts can ignite combustible material. A label must be on electrical equipment with the manufacturer's name and other descriptive markings, which clearly state the voltage and current wattage. This information is used to determine electrical loads.

To prevent the hazards of mismatched wire sizes electrical circuit breakers must open automatically if excess current from overloading occurs. To ensure this happens workers must use the proper fuse or circuit breaker.

Replace damaged cords; however, a licensed electrician can replace the plug. Cords are usually cheap to replace, so it is not necessary to use a cord that has the outer cover:

- Cut

- Burned
- Repaired with tape.
- Cracked or damaged
- Damaged by staples, nails, or fasteners.

Extension or flexible cords should be watched when run through holes in walls, ceilings, floors, doorways, windows, or similar openings. They should never be hidden in walls, ceilings floors, conduit, or other raceways.

Workers should not be permitted to work in proximity of any part of an electric power circuit where the worker might contact the electrical power circuit unless the circuit is de-energized, grounded, or guarded by insulation to prevent the worker from contacting the energized parts.

In situation where this cannot be done extra precautions must be taken when working live. There are specific items of PPE that can protect workers from electrical hazards. Proper foot protection includes a solid, well-built shoe that is nonconductive. Workers can use rubber insulated gloves, hoods, sleeves, matting, and blankets. Hard hats should be insulated and nonconductive.

Grounding and Bonding

"Grounding is conductive surroundings of an electrical system or circuit, usually assumed to be earth. Grounding is also the connecting of all enclosures of an electrical installation together and to the grounding point at the source of the system. An established grounding connection prevents accidental faults from occurring to energized conductor enclosures, the ground fault current will follow this established path. Effective grounding will open the circuit breaker or fuse, preventing dangerous voltages on the enclosures" (Fanning, 2003).

Grounding is important when working with power tools. That is because grounding creates a low path of resistance from a tool to the earth to disperse unwanted current. When a short occurs, electrical current flows to the ground protecting the worker; however, if the ground is broken or disabled the electricity can go through the worker to the ground. This can result in injuries or death of a worker. Tools are made with double insulation to protect the worker from being the ground. More information is in chapter 4 on power tools.

If a device is plugged into an outlet with an energized ground or into an extension cord with broken wire electricity can run from the tool through the worker to the ground. The Ground-Fault Circuit Interrupter or GFCI is a device that detects the difference in current between the energized and neutral wires. If the ground is detected the GFCI shuts off the electricity within 1/40th of a second. This prevents the electric current from going through the workers to the ground. GFCIs are required on construction sites. Contrary to popular belief the GFCI protects the worker not the tool or piece of equipment.

In lieu of GFCI, companies can use an assured equipment ground conductor program. This is not the best option because it can be a lot of work. This program covers all cord sets, receptacles not part of a building or structure, or equipment connected by plug and cord. The program requires specific procedures that include having a competent person to implement the program and that the program provides visual inspections of equipment connected by cords and plugs for damage. Damaged equipment must be tagged and disposed of or repaired. All other equipment passing the inspection is tagged.

When working around flammable material, containers should also be grounded to prevent static electricity. The receiving containers should also be bonded to the container releasing flammable liquids. This can be done by touching the metal nozzle of the first container to the second container, or use a bonding cable (Fanning, 2003).

Control of Energy Sources

Control of Energy Sources is often referred to as Lockout/Tagout or LOTO. This means that equipment or circuits are de-energized or made inoperative by using a tag, lock or another method that prevents another worker from re-energizing the equipment or circuit manually or accidentally while someone else is working on it. LOTO is an in-depth program that requires far more than can be discuss in a short book; however, more information is contained in chapter 12.

The employer must have a written document that outlines the LOTO program, which must be reviewed annually and update when changes occur. All operations must be controlled where energy sources are present during maintenance operations. Employers must also have procedures in place to provide for tagging or locking-out of the power source before maintenance begins. Follow this by training workers on the specific duties they hold during maintenance operations. This training must include the importance of not energizing a machine that is tagged or locked-out. Companies must remember to train supervisors on their duties too. They must then make proper tagging and lockout material and equipment available to workers. Finally, an evaluation program must be in place to check the effectiveness of the LOTO program.

Summary

Electricity is a necessary part of any construction project. Unfortunately, dangers come with using temporary wiring and equipment while installing the permanent and much safer electrical wiring and equipment that is part of the construction project. The hazards of working with electricity are exacerbated when power is used in extreme weather like rain and snow. The good news is that these risks can be controlled so that workers face minimal risk. The OSHA Publication 3075 titled "Controlling Electrical Hazards" is an excellent publication that provides an overview of basic electrical safety and can be obtained from http://www.osha.gov/Publications/osha3075.pdf. OSHA Publication 3007 titled "Ground-Fault Protection on Construction Sites" is an excellent resource for GFCI use. This publication provides basic information on GFCI for safe use of portable tools and electrical cords. It is available at http://www.osha.gov/Publications/osha3007.pdf. There is also a short fact sheet on LOTO that can be accessed at https://www.osha.gov/OshDoc/data_General_Facts/factsheet-lockout-tagout.pdf.

Chapter 7–Scaffolding

Falls are the most dangerous hazard that workers are exposed to when working on scaffolding. OSHA 29 Code of Federal Regulation, Part 1926.451, subpart L is the primary source for these types of risks. Scaffolding is described as an elevated temporary working platform. These platforms enable workers to work at a height above ground from safe and secure working platforms. There are three basic types of scaffolding:

- Suspended Scaffolds–platforms suspended by ropes or other non-rigid, overhead supports.
- Supported Scaffolds–platforms supported by rigid, load-bearing members, such as poles, legs, frames, or outriggers.
- Aerial Lifts–equipment or vehicle that remains on the ground and elevates a work platform to the height of the work.

Hazards

Supported scaffolding is built up to the height of the work. At these heights, the scaffolding can tip and fall because if it becomes top heavy. Furthermore, the scaffold can tip if the one of legs sinks into the surface or ground.

Workers can slip and fall from ladders that are used to climb the scaffolding. In some cases, the ladder does not reach to the ground forcing the worker to jump from the ladder to the ground creating the risk of injury from the landing. There is also slip and fall hazards when using ramps and walkways to climb the scaffolding. If workers use cross braces to climb up a platform the risk of slipping off and falling is significantly increased.

The ropes and cables of suspension scaffolds can fail resulting in the scaffolding falling to the ground. Also, with this type of scaffolding the platform can sway in the wind causing damage to the building and scaffolding as well as injuries to the workers.

In some cases when scaffolding is built around overhead electrical wiring it leans or tips into the wiring that can form a ground resulting in damage and injuries to the workers on the scaffolding from electrical shock.

Bad planking on the platform can give way or the worker could slip or trip and fall off an unguarded end of the platform, and the platform itself could fail falling to the ground. There is also danger from tools and material falling off the platform onto employees and public below the scaffolding.

Controlling Hazards

The primary focus for controlling hazards on scaffolding is to prevent falls. If a worker can fall more than 10 feet from scaffolding, he or she must be protected by Guardrails and a Personal Fall Arrest System (PFAS). Guard Rails for scaffolding are the same as those used on stairs:

- Install along open sides & ends.
- Toe Boards at least 3-1/2 inches high
- Mid Rails halfway between top rail and platform
- Top rails should be 39 to 45 inches above surface.
- Front edge of platforms no more than 14 inches from the work

The secondary focus is on preventing objects from falling onto workers and the public below the scaffolding. Of course, the toe board on the scaffolding helps prevent objects from falling. Wearing hard hats can protect workers below the scaffolding. Barricades and screen panels can block off the area below scaffolding to entry. Lastly, a canopy can be built, or a net erected below the scaffolding that will contain fallen objects.

Worker Training

There are two types of workers who need to train, erectors and users. For workers who will build or dismantle the scaffolding the employer must provide training in erecting, disassembling, moving, operating, repairing, maintaining, or inspecting scaffolding. The learning objective is for workers to recognize the hazards and the correct procedures to erect or disassemble the scaffolding.

Employers must also train workers that work on the scaffolding, but not erect or disassemble. For these workers, the training includes the hazards of working on a scaffold and procedures to control those risks. The training should include:

- Scaffolding load capacities
- Proper use of the scaffolding
- Nature of electrical, fall, and falling object hazards.
- How to deal with electrical hazards and fall protection systems.

For scaffolding workers, the training should also include highlighting the danger of remaining on the scaffold while it is being moved unless the surface is level, the height to base ratio is two to one, and outriggers installed on both sides of scaffolds. All these requirements prevent tip over as the scaffolding is being moved.

If the worker will be on suspension scaffolding, the worker must be trained to protect suspension ropes from heat and acid. They should be trained not to work on scaffolding when snow or ice covers the platforms or during storms or high winds. When an operator is using overhand bricklaying from supported scaffolding, they must be taught that a guardrail or PFAS is required on all sides except the side where the work is being done.

For both categories of workers all training should be documented, and performance examinations used to ensure workers have obtained the knowledge, skill, or ability desired. Training should be performed annually or when equipment or procedures change.

Competent Person

Scaffolding is one of those areas where the OSHA requires that a competent person be appointed and trained. This worker must be capable of:

- Training workers to recognize hazards.
- Selecting qualified workers to conduct work.
- Identifying and promptly correcting hazards
- Stopping work if necessary, to prevent an accident.
- Determining if it is safe to work during storms or high winds.

The competent person must be on site to supervise the scaffolding use. He or she must evaluate connections to ensure the supporting surfaces can support the load and inspect ropes for defects before each shift.

Scaffolding Erection

Scaffolding is erected on the site from pieces that go together to create a platform for workers, equipment, and material at the height of the work. Essential elements of safe scaffolding construction include using appropriate construction methods, proper access, and platforms upon which the workers can safely work. To ensure scaffolding will not fall over remember that the height of the scaffold should not be more than four times its minimum base dimension unless guy wires, ties, or braces are used.

The structure must be erected on stable and level ground with wheels locked and braced. The component pieces must match and be of the same type. They must be able to support its weight and four times the maximum load expected on them. The scaffold should be fully planked between front upright and guardrail support with no more than one inch gap. The width of the platform must be at least 18 inches. There cannot be a large gap in the front edge of the platform. The platform should be made of scaffold grade unpainted wood. The platform ends must be extended over its support by at least six inches, unless cleated or otherwise restrained by hooks.

If a worker can fall more than 10 feet from scaffolding, he or she must be protected by guardrails or a PFAS. PFASs must have anchors independent of the scaffold support system. Electric shock or electrocution is a serious consideration when erecting scaffolding near overhead power lines. It is essential to check the clearance listed in the standard. Support devices must rest on surfaces that can support four times the load.

Scaffolds can only be erected, moved, dismantled, or altered under the supervision of a competent person. The competent person selects and directs workers and determines the need for fall protection. After the scaffold has been erected a competent person inspects it for visible defects before each shift and after any alterations.

Scaffolding Usage

The competent person must be always on site the scaffolding is used. Before each shift, the connections must be evaluated to ensure the supporting surfaces can support the load, and ropes must be inspected for defect.

The safety measures put in place for the scaffolding must be enforced during use. The most important safety measure is that only trained workers be allowed to work on scaffolding. Particular attention should be on how workers are accessing the scaffolding and the load placed on platforms. Guard rails and toe boards should remain in place. Workers below the scaffolding should wear hardhats and only enter the area when they must. The area below scaffolding should be barricaded to block off the area below scaffold to forbid entry into that area.

Scaffolding Removal

Scaffolding is dismantled on site from the top down by removing the items that were put together earlier. Suspended scaffolding is usually pulled up on to the roof and dismantled. The scaffolding should only be dismantled by workers who have been trained. Scaffolding should also only be dismantled if the Competent Person is present.

Essential elements of safe scaffolding dismantling include using appropriate scaffold dismantling techniques, proper scaffold access, and platform removal. If the height of the structure is more than four times its minimum base dimension and guys, ties, or braces were used temporary means must be used to hold the scaffolding erect while it is being dismantled.

A means needs to be present to lower platform planking or decking and the larger pieces of the scaffolding. Workers performing the dismantling are still exposed to fall hazards. When dismantling scaffolding near overhead power lines the danger of electric shock or electrocution hazards still exists. It is essential to maintain clearance distances listed in the standard. The competent person selects and directs dismantling workers and determines the feasibility of fall protection.

Summary

Scaffolding is a crucial part of any vertical construction project. It allows the workers to work at an elevation above the ground. It also provides in proximity tools and material that facilitates the work. Despite the dangers of falling from scaffolding or dropping objects onto workers below, working on scaffolding can be done safely. If the workers are trained and the rules and standards followed, jobs can be done efficiently and effectively. Some of the necessary information was provided in this chapter; however, remember that the competent person must be always on site to supervise the scaffolding. The OSHA Publication 3150, "A Guide to Scaffold Use in the Construction Industry" provides great information and can be obtained from http://www.osha.gov/Publications/osha3150.pdf.

Chapter 8–Fall Protection

Introduction

Falls are the leading cause of death in the construction industry. In addition to falls from scaffolding there are falls from open-sided floors or through floor openings. Falls from as little as six feet can cause serious lost-time accidents and even death. OSHA 29 CFR Part 1926, subpart M is the primary source for this type of hazards. To prevent falls guard open-sided floors and platforms six feet or more above ground.

Basics

If a worker can fall six feet or more onto a lower level, some form of fall protection must be provided. It is the employer's responsibility to guard the danger and implement a fall protection system to protect the worker. Where and when is fall protection required? The answer is that fall protection is required when workers are performing the following work:

- Roofing
- Bricklaying
- Excavating
- Wall Openings
- Walkways and Ramps
- Residential Construction
- Concrete Forms and Rebar
- Open Sides, Edges, and Holes

Prior to choosing fall protection for hazards on the construction site it is important to know it consists of three options:

- Guardrails
- Safety Nets

- Monitors
- Personal Fall Arrest Systems (PFAS)

In addition to falling off an object, a worker could step onto and break through skylights and other openings on roofs, floor, and above the ground. These skylights and openings must be protected if more than six feet above ground. On sites, holes in the floors must be covered completely and securely. If the cover is not available, the hole can be protected with a guardrail. It is necessary to use a PFAS when working on formwork or rebar. This is because there is a high risk of a worker falling onto the rebar and being impaled. To prevent this protruding rebar must be covered or capped. The cap referred to here is a plastic cap that has a surface larger than the rebar that is installed over the tip of rebar to prevent the impaling.

Guardrails are a handrail used to steady a worker while they work on ramps, runways, and other walkways where the employee can fall six feet or more to the ground. Guardrails consist of a top rail between 38 and 42 inches off the bottom surface, toe boards at least 3 ½ inches, and a mid-rail in between. If working on a roof, tether or restraints can be used to prevent workers from reaching the edge, thereby preventing a fall. Safety nets can be used to catch workers if they fall.

Safety nets should be hung as close as possible, but no more than 30 feet below the work area. Falls of more than 30 feet can result in workers injured by landing in the net. As noted earlier, it may be better to install a safety net below the workers to catch them if they fall. In 29 CFR 1926.502, (c), (3) it states that "safety nets shall be installed with sufficient clearance under them to prevent contact with the surface or structures below when subjected to an impact force equal to the drop test specified in paragraph (c), (4) of this section. OSHA 29 CFR 1926.502, (c), (4) states "safety nets and their installations shall be capable of absorbing an impact force equal to that produced by the drop test specified in paragraph (c), (4), (i) of this section."

Monitors can also be used. These are fellow workers that watch the locations of workers and stop them from getting close to the edge. This is the least desirable of all the systems because it relies on a worker to pay attention. There is usually a line of flags placed several feet back from the edge to help the monitor identify when a worker gets too close to the edge.

PFAS is comprehensive fall prevention that consists of an anchorage, lifeline, and body harness. Anchorage points secure the worker to a fixed object. The PFAS is harnessed in the worker's upper back. If the worker falls, an arrest system slows and stops the falling worker before he or she strikes the ground. They must be independent of any platform anchorage and capable of supporting at least 5,000 pounds per person on the PFAS. Lifelines are ropes that can slow and holding the fallen worker. Back belts are never acceptable replacements for PFAS. A body belt is fastened at the waist and connected at the front waist of the worker. If used by a falling employee there is no arrest system to slow the worker and when the belt catches the worker would likely break their back. Body belts are used to hold a worker at elevation and should not be expected to do anything else.

If an employee falls and is saved by a PFAS or a net, he or she will need to be rescued. You cannot simply pick the person up; the human body is dead weight at this point and would take a great deal of effort to pull up on a rope. An emergency preparedness plan must be in place that includes procedures for obtaining help from local emergency authorities like the fire department or emergency medical services. Early coordination is required between the construction company and the local authorities to ensure that the local authorities can rescue a worker after a fall and in fact will respond. Some emergency organizations are not capable of performing a fall rescue and others will not perform the rescue because of legal restrictions. If an emergency organization is found that can and will help, it is important to invite them for a site visit to ensure the emergency personnel are familiar with the site before an emergency occurs.

Training

The best person to prevent a fall is the worker; however, they must know what to do to keep from falling. Employers must let the workers know they want them to speak up when other workers are not using fall protection but should. It is the company that must provide fall protection training. The training is to teach the worker how to recognize and minimize risks. The training must include fall hazards, protection systems, and fall protection devices. The employer must have a competent person appointed to oversee the preparation and use of fall protection. The competent person must have training that will prepare them to fulfill the duties as the competent person on fall protection project. Employers must also provide training to workers who will assist in the use of fall protection. Workers must be trained again if the conditions change or workers demonstrate behavior that indicates they are not using proper procedures.

In all cases, it is important to document the content of the training as well as dates and times of training. A performance examination should also be used to verify the desired learning took place. The best practice is to have workers and trainers sign a roster to certify the training. Keep all training records for five years.

Summary

The costs associated with a fall can break a company and ruin the lives of workers and their families. Fall protection systems and work practices must be in place before workers start to work six feet or more above ground level to prevent falls. There are alternatives that can be used. Workers can perform work at ground level with prefabricated items on the ground and lifting them into place with a crane. This can reduce the time working at an elevation, which reduces the risk. A lift can also be used to raise workers to the work area. Whatever method is used to control the hazards this is time well spent. OSHA has developed an entire webpage to address fall protection that is a good place to find information. This information can be obtained from OSHA at https://www.osha.gov/doc/outreachtraining/htmlfiles/subpartm.html.

Chapter 9–Excavations, Trenching, and Shoring

Another hazardous construction operation is excavating. A collapsing trench is the primary hazard caused by dirt falling back into the trench trapping the worker, suffocating him or her. OSHA 29 CFR Part 1926, Subpart P, titled Excavation, Trenching, and Shoring and it is the primary source for these hazards. Workers are protected from collapsing excavations by using sloping, shielding, and shoring. It is important to start with a few definitions:

- Excavation – man-made trench or cut in the ground created by removing earth.
- Trench – narrow excavation with its width less than its depth
- Sloping – trench with the opening wider at the top than at the bottom with a gentle slope from the top to bottom. Looked at from the front it resembles a flat-bottomed V.
- Benching – trench with the opening wider at the top than at the bottom with stair like benches from top to bottom. Looked at from the front it resembles a stairway on each side.
- Shielding – device that protects workers set into the trench to keep walls open.
- Shoring – means of protecting workers by holding walls of a trench open using sheeting, posts, struts, and wales.

Part I Excavation and Trenching

Excavation hazards and risks to workers include:

- Falls
- Cave-in or collapse
- Accumulation of water
- Exposure to toxic fumes
- Limited access and egress

- Oxygen deficient atmosphere
- Mobile equipment working at or near side of the trench.

Falls can occur in and around the excavation. Walking over loose piles of soil can cause slips and falls that cause a worker to fall into the trench resulting in severe injuries. This hazard can be reduced by cutting walkways through the piles of soil so that workers do not have to climb over them.

Water from rain or run-off can destabilize the soil and cause a cave-in, collapse, or mudslide. Water bubbling up from the bottom of the pit creates additional risks. Inspections should be conducted of trenches after rain and not opened until hazards are controlled. Pumps can be used to remove water from the excavation reducing the risks.

Some excavations can have low amounts of oxygen in the environment. Workers need an oxygen level between 19.5% and 23.5%. If the oxygen level is below 19.5%, the worker can experience symptoms that can lead to death. This danger can be reduced by workers wearing a respirator that provides oxygen. If the oxygen content gets above 23.5% it also has detrimental effects on the worker and now presents an explosion hazard. A spark created by a tool can cause the oxygen to explode. In some cases, the use of forced air can also reduce the risk of explosion. The bottom line is that if the oxygen level is not at 19.5%, someone needs to find out why.

Excavations may also contain toxic fumes. Meters can be used to check the atmosphere before workers enter the channel. The check is done before the start of work in the trench and periodically if glues, solvents, or electric generators might introduce toxic fumes into the trench. In most cases the use of forced air can reduce the toxic fumes. In some cases, this danger can only be reduced by the workers wearing a respirator.

If the trench is more than four feet deep ladders must be provided to give workers a way to get down in to and out of the trench. The ladder needs to be placed every 25 feet in the trench to reduce the amount of distance a worker will have to travel to get out in an emergency. These ladders must be secured and extend above the top edge of the trench by three feet. Do not use metal ladders if electric utilities are present.

The most frequently occurring hazard for any excavation or trench is that a cave-in or collapse will occur. The weight of exposed earth at or near the edge or the weight of the soil in the sides can cause the dirt to fall into the trench burying a worker. Collapses can happen without warning and catch the workers in the trench off guard. Under the dirt, the worker can suffocate if not rescued. In most cases, workers cannot dig themselves out. Effort by other workers can take time and help may not reach the trap workers before they die. Recent rain can increase the weight of soil on both sides causing a cave-in.

The obvious protective measure is to dig a trench so that the sides are sloped or benched. Either method prevents the walls from caving-in. Sloping makes the trench wider at the top than at the bottom. The slope reduces the weight of the soil and the angle prevents soil from falling into the trench. Benching is similar except the sloping sides are benched that involves cutting steps into them. The benches serve the same purpose as the slope. In either case, the amount of slope or bench is determined by the soil type and the depth of the channel. Not all workers know how to determine the needed slope, but equipment operators or engineers usually have the knowledge of soil mechanics and can determine the amount of slope or bench needed. Information from the OSHA website is shown in table five that shows the slope or bench. This information should only be used by someone qualified in trenching and excavating.

The soil pile is loose dirt dug out of the trench stored on one or both sides of the trench. If not stored properly the soil can easily slide back down into the trench. Heavy or prolonged rain can also cause the soil to become muddy and hasten the slide back down into the trench. A protective measure is to keep the soil piles back two feet or more from the edge of the trench. Placing the soil removed from the channel on one side can reduce the risk while making the dirt readily available for putting back into the trench.

Soil Type	Height/Depth Ration	Slope Angle
Stable Rock (granite or sandstone)	Vertical	90°
Type A (clay)	¾:1	53°
Type B (gravel, silt)	1:1	45°
Type C (sand)	1 ½:1	34°
Type A (short term) (For a maximum excavation depth of 12 feet)	½:1	63°

Table 5–Allowable Slopes (OSHA, 2014)

It is important to keep motor vehicles back away from the sides of the trench. A backhoe or excavator may be used to lower pipe or other material into the trench, but if the vehicle is too close to the edge the vehicle's weight can cause a collapse.

Working in the trench creates its own hazards. Lowering equipment and material into the trench is often done. To reduce the risk workers should never work under a load. If the weight is dropped or swung, a worker underneath could be injured or killed by the load. Workers can also slip, trip, and fall when stepping over equipment and material in the trench. To reduce these hazards, create a walkway in the channel free of obstructions.

Once the trench is made safe, the competent person should begin regular inspections of it. Many things can happen overnight or over the weekend that create hazards. Rain is probably the most common one. Removing ladders by thieves can hamper getting in and out. Oxygen levels can drop, or toxic fumes can occur. A quick daily inspection can prevent or eliminate these hazards.

There are several requirements for protective systems with respect to the trench. It is necessary to design sloping and benching correctly, and that workers appropriately handle materials and equipment around the trench. OSHA has a web page that is focused on trenching and excavation that provides accurate help. It can be found at https://www.osha.gov/SLTC/trenchingexcavation/construction.html.

Part II Shoring

Protective systems are used when sloping or benching cannot be done or in addition to them. These systems must be bought and used in accordance with manufacturer's directions or designed and certified by a licensed professional engineer. Protective systems must be able to resist all expected loads to the trench that were identified by an analysis. Shoring is a structure that supports the sides of the trench and protects against cave-ins. Protective systems can be used to prevent cave-ins and protect workers. The following systems can be used:

- Shield systems
- Support systems
- Timber shoring system.
- Mechanical shoring system
- Metal hydraulic shoring system
- Portable shields (trench boxes and trench shields)
- Permanent shields (trench boxes and trench shields)

The factors that should be considered prior to using shields include:

- Soil Classification
- Depth of excavation
- Framework to work in
- Changes due to weather.
- Water content of the soil
- Support for excavation walls
- Use of wales, cross braces, and uprights.

Workers can build shoring in the trench. Boards are placed against the wall next to each other. Uprights are then placed against the boards to hold them in place and cross braces are placed between the uprights to hold them apart. The whole system pushes boards against the walls of the trench preventing a cave-in. The members must all be properly sized depending on the dangers of a cave-in. Shoring built in the trench can be time-consuming. It is also easy to lose one of the many pieces.

A manufactured system is built to be lowered into the trench and adjusted by a standard screw to fit the side walls will save time. Systems are manufactured and can be purchased at a reasonable price. These can be installed in each area where workers are in the trench. These can even be placed one on top of the other for deep trenches. Hydraulic jacks can also be purchased to replace the standard screw mechanism. The hydraulic pressure is used to keep the sides in place, and a trench pin placed in the mechanism creates safety in case of hydraulic failure.

Whatever method is used, or equipment purchased it must be kept in good working order. This is done through correcting deficiencies found during daily inspections. Always make sure a professional engineer approves the equipment for use and a competent person is trained and on site while excavations being worked.

Employee Training

In addition to being trained as a plumber, electrician, or any other trade workers are trained to work safely in an excavation. The training should include basic information about the hazards of working in the trench, how to control risks, how to respond to an emergency, and the proper safety clothing and equipment workers should wear.

Much of this information can be found in the 10- or 30-hour OSHA Construction Safety Course. The training company can adjust the schedule to give more information on working in the excavation and reduce the information in other areas as needed. This training should be repeated every three years. Organization uses trench boxes or shoring workers must provide specific training on the worker's responsibilities with respect to these devices as well as installation and removal procedures.

A competent person is usually a plumber, electrician, or other trade who has been designated as the competent person for excavation. He or she should have extensive work experience with excavations. The competent person should also complete the 10 or 30 OSHA Construction Safety Course every three years. Furthermore, the competent person must complete specific training to provide advanced training for excavation, trenching, and shoring. A quick search of the World Wide Web can provide information on companies that provide this training.

Personal Protective Equipment (PPE)

The employer must provide protective clothing and equipment required by the actual work done in the excavation. Most workers will need gloves and hardhats. Workers welding in the trench need to wear all the PPE for the hazards of welding. Respirators might be needed for trenches with toxic fumes from sewage.

Summary

With excavations being recognized as one of the most hazardous construction operations, it is necessary to identify the risks and take steps to prevent them from causing injuries and illnesses for workers. The primary hazard of excavation is a cave-in. It is the employer's responsibility to protect workers from cave-in. The best way to start is by identifying hazards, training all workers, and using a competent person. OSHA Publication 2226 is an excellent primer for these hazards. The booklet is in a question-and-answer format and answers all the basic questions about working in excavations safely. You can download it free from OSHA (OSHA Booklet, 2014).

Chapter 10-Stairs and Ladders

Construction workers fall from stairs and ladders nearly every day. Most of these falls do not result in injury; however, many do. A worker can die from a fall of less than six feet. It is important for each construction site to observe good practices for stairs and ladders to prevent falls. OSHA 29 CFR, Part 1926, Subpart X: Stairways and Ladders is the primary source for this type of hazard. On construction sites, there must be a stairway or ladder whenever there is a 19-inch increase in elevation. The stairway or ladder also provides a point of access for workers. All access points must be kept clear of obstructions.

Stairs

Falls are the primary hazard associated with stairs. These falls cause injuries and fatalities among construction workers. About half the injuries from falls require that the worker need time away from work. Many workers sustain serious injuries from falling as little as six feet.

Stairs consist of risers, treads, landings, and hand and stair rails. The riser is the vertical part of the step, and the tread is the horizontal part of the step. Stairs should be installed at an angle between 30 and 50 degrees, and they should have uniform riser height and tread depth, with less than a ¼-inch of variation.

The handrail is a single rail fastened to the wall between 38 and 42 inches above the tread for the full length of the stairway to provide employees with a handhold for support. Stair rails are vertical barriers installed on the open sides of the stairs. Stair rails consist of a handrail, mid-rail, and uprights spaced along the length. There must be one handrail installed when stairs consist of four or more risers or higher than 30 inches exists. Handrails and top rails are used to steady workers and must be able to withstand the force of 200 pounds, applied along any point of the top edge.

If the construction site has landings with open sides, a standard 42-inch guardrail system must be installed on those open sides. A landing must be provided every 12 feet from the ground. Stairway landings must be at least 30 inches deep and 22 inches wide. Where doors or gates open directly on a stairway, a platform must be provided that extends at least 20 inches beyond the swing of the door. Guardrail systems may also be needed on platforms with a swinging door to protect from potential falls of six feet or more.

These hazards are exacerbated by using temporary structures to get to the work areas above the ground. Temporary stairways have permanent treads and landings that are to be filled in later. These are concrete forms or pans that are filled with concrete after the stairs set in place. Metal pan landings and treads are secured in place before filling with concrete.

Slippery stairs are a common danger caused by dirt, snow, or water. These conditions should be corrected as they are found. Stairway parts also must be kept free of projections that can cause injuries or snag clothing. It is important to build stairs and platforms to standard and fasten them properly so as not to weaken or fail. Stairs and platforms should always be kept clean of obstructions. Neither is a good location to store tools or material.

Ladders

There are many types of ladders for many different purposes. There are portable and fixed ladders. There are portable ladders that support themselves and others that do not support themselves. There are fixed ladders attached to buildings, towers, and antennas.

Non-self-supporting ladders are ladders that lean against a wall or other support. To get the correct ladder the weight of the worker and the weight of everything carried up and down the ladder must be determined. Furthermore, a ladder must support four times the maximum expected load. Manufacturers rate their ladders so users will know how much weight the ladder can hold. According to the Ladder

Safety Institute, there are five categories of ladder Duty Ratings (American, 2014):

- Type IAA (Extra Heavy Duty) 375 pounds
- Type IA (Extra Heavy Duty) 300 pounds
- Type I (Heavy Duty) 250 pounds
- Type II (Medium Duty) 225 pounds
- Type III (Light Duty) 200 pounds

Ladders are made of wood, metal, and fiberglass. When working with electricity, a non-metallic ladder must be used to reduce the risk of electric shock.

Non-self-supporting ladders are leaned against a wall or support. When using a ladder for access to the upper landing surface, the side rails must extend at least three feet above the upper landing surface. In addition, the ladder is placed at an angle with the foot away from the wall the distance ¼ of the ladder length. A quick example: when using a 20-foot ladder the bottom feet are positioned five feet out from the wall the ladder leans against. This formula is ¼ of 20 feet or five feet. The worker must add the ¼ of the ladder length and the additional three feet to the length to determine total ladder height.

On construction sites, Double Cleated Ladders are used. This ladder has double-cleats with a center rail. This type can be two or more ladders joined together. There is only one place to use this type of ladder, when it is the only way for 25 or more employees to enter or exit the work simultaneously.

Controlling Hazards

There are many things to do to prevent hazards from becoming a reality. It begins with selecting the right ladder for the job. Using ladders for their intended purpose is also important, workers should never:

- Tie ladders together to make longer sections.
- Place ladders on top of things to gain extra height.
- Paint a covering on a wood ladder that prevents seeing damage.

- Use the top step of a stepladder as a step.
- Use cross bracing on the rear of a stepladder for climbing.

Because portable ladders are moved and transported, it is imperative to keep them in a safe condition and good working order. The rungs, cleats, and steps must be level and uniformly spaced at 10 to 14 inches apart. Side rails must be at least 11½ inches apart. A competent person should inspect ladders before each use. Those with visible defects should be removed from service and label with "Do Not Use" installed.

Ladders should only be used on stable and level surfaces unless secured. To secure a ladder, attach it to a fixed object. This will prevent it from accidentally moving due to the work being done on the ladder. The area around the top and bottom of the ladder must be kept clear. The area at the bottom of the ladder should also be kept free of slipping hazards.

When climbing the ladder workers should face the ladder whether going up or down. Workers should use three points of the contract while climbing and not carry objects or loads that could cause them to lose balance. This always means two hands and one foot or two feet and one hand on the ladder.

If the fixed ladder is longer than 24 feet, it must be equipped with a ladder safety device or self-retracting lifeline. This type of ladder also requires a rest platform every 150 feet or less, a cage or well, and multiple ladder sections not more than 50 feet in height.

Workers must be trained to use ladders safely. This training should be done by a competent person and provided before a worker uses a ladder. The training should help workers understand:

- What causes falls?
- The proper selection, use, placement, and care of ladders
- How to properly erect, maintain, and disassemble a ladder.
- The maximum intended load-carrying capacities of ladders

Knowing this information helps workers identify risks and the steps necessary to prevent hazards from causing an accident.

Summary

Stairs and ladders allow construction workers to work at heights, which is essential to their work because workers need to build a building at all levels. This puts workers at risk of falls. A fall from even six feet can result in the death of a worker. Lower falls can result in injuries and incapacitation. Stairs and ladders are tools to work with and must be built and used properly to eliminate hazards. Workers must be trained to ensure they use ladders correctly. This chapter outlines the things that workers can do to prevent falls. Additional information can be found in OSHA Publication 3124, titled "Stairways and Ladders." This booklet that explains OSHA's requirements for stairways and ladders can be found at http://www.osha.gov/Publications/osha3124.pdf.

Chapter 11–Confined Space

Confined Spaces are something that a worker gets into to install, repair, or maintain something; however, the space is not designed for the worker to remain there. Despite the spaces being big enough for a worker to enter they have limited or restricted means of entry and exit. The opening to these areas is usually smaller than a door and often located above ground or requires the operator to enter from the top or even the bottom (Safety, 2014). The 29 Code of Federal Regulation, Part 1910, Subpart 146, Permit Required Confined Spaces is the primary source for these hazards. It is important to note that this is a general industry standard and not a construction standard, which is why it is found in 1910 rather than 1926. There have been discussions about developing a confined space rule for construction, but until one is approved the general industry standard applies (Federal Register, 2014).

It is good to begin with some definitions before going too far with this chapter:

- Attendant – worker that remains outside of the confined space, but with the entrant and is prepared to assist the worker with exiting the confined space and calling for emergency assistance if needed.
- Confined Space – space large enough for a worker to enter, with limited means of access and egress, and not meant for continuous occupancy.
- Entrant – the worker who enters a confined space.
- Permit required Confined Space – a confined space with a hazard to life and property that requires a permit to enter.
- Safety Representative - qualified person that evaluates hazards and equipment, and issues the Confined Space Entry Permit (Safety, 2014).

Basic Confined Space Information

It is not hard to imagine that there are hazards to entering and working in a confined space. The first and perhaps most important is that there be enough oxygen to support the worker in the space. A human worker needs between 19.5% and 23% oxygen concentration in the air they breathe. Too little oxygen (less than 19.5%) is called oxygen deficiency and too much oxygen (more than 23.5%) is called oxygen rich. When the air is tested in confined space it should have an oxygen concentration within that range. Workers are in danger if the oxygen content is outside of the range. Also, with an oxygen-rich environment present any combustible materials may ignite or even explode. It is important to check for oxygen deficiency throughout the working period.

The worker in the space consumes oxygen as they breathe replacing oxygen with carbon dioxide. Also, in the space might be a gas that is heavier than oxygen that pushes the oxygen out of the space. The worker might also add toxic fumes to the space from using glues or cleaners. The worker might also add carbon-monoxide from welders, torches, and combustion engines. These pieces of equipment also use oxygen and can affect the amount of oxygen left in the area. The area must be checked periodically throughout the work period to make sure the oxygen level remains within the range that the worker needs.

There are also several toxic fumes and gases that can exist in a confined space. One example is a manhole, which is usually covered for long periods of time. In this manhole naturally occurring toxins, such as hydrogen sulfide can accumulate creating a danger. Flammable gases may also accumulate in the manhole. Toxic and flammable materials may exist in sanitary and storm sewers that are now in the manhole. There may also be tanks with leaks that have migrated underground causing seepage into manholes.

Another hazard that occurs within confined spaces is engulfment, which occurs when a sudden release of fluid solids, such as granulated salt, sugar, or sand, liquids, and dense gases cover the worker inside a silo, underground tank, or pit. These incidents can be deadly. The worker is covered with the material and can suffocate.

Working in confined spaces often involves complex exposures that include mechanical, electrical, pressure and chemical hazards. For example, while working in a utility tunnel the worker is faced with extreme heat, moist working conditions, and deep in the tunnel the oxygen concentration is too low. All these risks need addressed before the worker should be allowed to enter and work.

Hazard Identification

Confined spaces must be labeled. If a work crew is performing maintenance or alterations, they should see labels on confined spaces as a warning; however, it is difficult to identify every single space. That is why it is always wise to make annual checks to identify any new or missed confined spaces. Once the confined spaces have been identified workers should conduct a hazard assessment of each space. If the space hazards, it is labeled a confined space. When hazards are identified the confined space is identified as a permit required confined space.

A permit is required to work in these spaces because permit confined spaces have hazards life and property. Company's employing workers in permit required confined space must have a Confined Space Entry Permit Program. That program has the processes in place to provide the worker with a signed Confined Space Entry Permit prior to entering the space.

Prior to getting a permit a safety representative will visit the space to evaluate conditions under which the entry is made. The safety representative examines the space for concentration of oxygen. They follow this by testing for hazardous concentrations of known harmful substances, such as hydrogen sulfide, carbon monoxide, and flammable liquid or gas. A single meter is normally used that measures all these types of gases. It is important to take measurements at the point of entry and intervals within the space. For example, if the work is performed in a manhole tests should be made at one-foot intervals from the top to the bottom of the hole. The readings are noted on the worksheet that is prepared prior to filling out the permit.

If the concentration of oxygen is not correct or if toxic fumes or gases are present ventilating the confined space is usually the first step to increasing oxygen concentrations and removing toxic fumes and gases before entrance. While ventilating tests are retaken to see if readings have improved. If they have the worker may be able to enter with the ventilation providing continuous exchanges of air while a worker is in the space. Ventilating is usually done with outside device that forces air into the space supplying fresh air and diluting or removing contaminants. If ventilation does not work and the concentration of oxygen and contaminants remain at harmful levels, respirators may be worn to supply safe air to the worker. To improve the oxygen concentration respirators can have oxygen sent to the wearer.

Hazards from adjacent operations or processes may leak or bleed over to the space and should be exposed and checked. For each hazard, the safety representative identifies a control measure to eliminate or control the hazard. These measures are noted on the worksheet and the Confined Space Entry Permit.

In some cases; people will be walking, driving, or riding bicycles near the confined space. Barriers should be placed around the area to prevent these people falling into the opening. Most common barriers are the railing system with cones that most cities use to identify the manhole when the cover is removed. Warning signs stating the hazard and actions required should be placed where people can see them. When work is completed, or workers have left the job all holes and openings must be closed or guarded.

Hazard Control and Abatement

Each hazard should be identified and controlled or eliminated. Hazards that cannot be increase the risk for workers involved and must be accepted by a senior person who will approve the work to proceed.

There are three basic kinds of control measures. First the hazard should be engineered out. Next is for workers to wear PPE. And lastly administrative controls are used. These controls consist of training, reassigning workers, and developing standard operating procedures.

PPE can control many hazards. Workers should wear safety glasses to prevent eye injuries from flying sparks, liquids, and welding lights. Wearing hardhats may also be necessary to prevent head injuries from items falling on the head or the worker striking their head against an object. If oxygen is not in the proper concentration, oxygen providing respirators can be worn. If toxic fumes or gases are present, then a filtered or air providing respirator may be used. When working in containers that previously held flammable liquids workers can wear coveralls and booties to prevent sparks. If objects are present that can drop on the foot, the operator can wear safety shoes. If hands can be injured by sharp or rough objects, hot objects, corrosive chemicals, or other toxic chemicals protective gloves can be worn.

Electrical hazards also need to be controlled. Many confined spaces are damp and my not provide a good earth ground for portable tools. Using GFCI's on all portable electric tools reduces the danger of a shock. They have unique models for use on construction sites that are more rugged than standard models. There may also be energy sources in the confined space that will need locked out or tagged out. This may also mean that some energy sources on adjacent equipment and processes may need to be locked out or tagged out too.

Since conditions can and often do change while work is being done in confined spaces, it is important for the attendant to monitor for oxygen and toxic and combustible gases throughout the working period. It is also common for workers in the space to wear a personal monitor. This monitor has an alarm that sounds when the oxygen level goes down, or the device detects dangerous levels of toxic and hazardous fumes. The worker immediately leaves the confined space when the monitor alarm activates. The attendant outside the confined space must make sure the worker in a confined space leaves when an alarm sound.

If best efforts to identify and control hazards did not work, and an accident or an emergency happens, the company's emergency response procedures for confined space are initiated.

Rescue Methods and Procedures

Once the hazards and control measure are identified it is important to ensure a rescue method is in place if something goes wrong. Employers must have procedures in place that are known by all workers to rescue workers in a confined space. This is defined as a means of emergency rescue readily available to the attendant for emergency extrication of workers in the confined space. All workers on the site must know the emergency rescue procedures.

There are a variety of safe methods for workers to exit the confined space under normal circumstances, as well as an emergency extrication. The most used method is a tripod with hoist, lifeline, and full body harness. Allowing the worker to be lowered into the confined space and if something happens the worker can be pulled out using the rope and hoist. Workers normally use ladders for ordinary entry and exit while a rope and hoist are available for emergency rescue.

The attendant should remain in constant contact with the workers inside the confined space. If conditions develop that require extrication, and the worker cannot get out of the confined space on their own, the attendant needs to call for emergency assistance. All workers who may serve as attendants must know the phone numbers to call and a telephone to make the call. After calling for emergency assistance the attendant should remove the worker from the confined space using the tripod, hoist, and lifeline or another method. The attendant should never enter the confined space to rescue a worker. Fifty percent of workers who die in confined spaces are the attendant. Only equipped and trained emergency rescuers should enter confined spaces to make rescues.

Companies should contact emergency service providers to check if the service has the proper equipment and will respond to a confined space emergency. It is unfortunate when the first time an emergency service hears from the company when an emergency is unfolding.

Permits

Permit Required Confined Spaces require a permit before any workers enter the space. The employer must have a Confined Space Entry Permit Program in place. This program consists of a written standard operating procedure, processes, and permit forms.

There is a lot of information that is on the entry permit. This information includes such things as the date, location, and the name of the confined space. The purpose of the entry, identified risks, and control measures are also included. The duration of the entry is added, and the permit is void after that time. The authorized entrants, attendants, and the site supervisor are also listed. Air testing results are listed on the permit and the person conducting the tests signs. If ventilation, isolation, flushing, or purging were done to remove or reduce risks is recorded. Lockout/tagout methods that are used must be identified. The name and phone numbers of rescue and emergency service provider is listed as well as the communication procedures that will be used. Any special equipment and procedures that will be used must be listed. Any PPE used as well as alarms and rescue equipment and procedures should be listed.

The supervisor signs the permit only after verifying pre-entry precautions are taken and the area is safe to enter. The permit is posted at entry point of confined space. The permit terminates when the task is completed or when new conditions exist that require a reevaluation.

Training and Education

Training is required for all workers who enter confined spaces, serve as attendants and those who will be rescue team members. This training must be conducted prior to the initial entry. Retraining is provided any time the duties of the workers change, the permit-program changes, and new hazards are present or introduced, or work performance indicates deficiencies.

Summary

Construction workers can find themselves working in and around confined spaces. It is critical for employers to have a plan in place to deal with these risks before any work is done. Confined space entry hazards can include insufficient concentration or too much oxygen, toxic substances, engulfment, flammable gases, combustible liquids, equipment related hazards, and conditions that change from nonhazardous to hazardous. Prior to conducting work in confined spaces, employers must develop standard procedures and ensure all workers are trained on the specific hazards of their work as well as general hazards and procedures for working in confined spaces. With a confined space entry program in place, workers can work in confined spaces safely. OSHA has created a short booklet that provides essential information for work in confined spaces. This booklet is found at: https://www.osha.gov/Publications/osha3138.pdf.

Chapter 12–Controlling Hazardous Energy Sources

There are dangers that exist in workplaces that are thought about after a worker is injured or killed. These hazards are known as Energy flows and come from machine, electrical, or equipment. If a worker is maintaining that machine, electrical devices, or equipment, he or she will usually shut the equipment off. What if another worker came along, did not see the worker performing maintenance and turns it back on? The worker is now in danger of that energy and probably does not even know it. What if the machine were disabled so that no one except the worker performing maintenance could turn it on? The danger would not exist. OSHA 29 Code of Federal Regulation (CFR), Part 1910, Subpart 147: The Control of Hazardous Energy is the primary source for these hazards. The secondary source for our purposes is CFR 1926, Subparts K and Q.

It is important to know that workers can die from injuries sustained when working with hazardous energy sources. Stored energy can kill. Stored energy can also allow heavy equipment to close crushing the worker or heat to be released through steam scalding the worker. There are five main causes of injuries that include accidentally restarting of equipment and failure to:

- Stop equipment.
- Disconnect from the power source.
- Clear work areas before restarting.
- Dissipate (bleed, neutralize) residual energy.

There are practices and procedures to disable machinery and equipment so that workers are protected from energy sources while performing services and maintenance activities. The intent is to prevent unexpected start-ups or releases of stored energy that could injure a worker. OSHA gives the opportunity to do this by one of two methods. The workers can lockout an energy source by deactivating it and locking it out with a padlock to prevent someone from reactivating it again. The second option is for the worker to tagout the source with something like a shoe tag; however, the workers can only use the tagout procedure if it provides equivalent protection to the lockout procedure. The acronym LOTO is used for this program to represent Lockout/Tagout. Whichever method used the worker will leave their name on the lockout or tagout device to let other workers know who has control of the device. The locks and tags remain in place until the work is complete.

Written Program

The employer must have a written plan in place prior to conducting any work on equipment powered by energy sources. The written plan must describe the roles and responsibilities of workers and supervisors, methods used, training requirements to be followed, and establishes the policy that only the worker that applies the lockout or tagout can remove the device. OSHA does allow an exception "When the authorized employee who applied the lockout or tagout device is not available for removing it that the device may be removed under the direction of the employer, provided that specific procedures and training for such removal have been developed, documented and incorporated into the employer's energy control program. The employer shall demonstrate that the specific procedure provides equivalent safety to the removal of the device by the authorized employee who applied it" (Exception, 2014).

In the written program it is necessary for the company to identify tools that are authorized to be used by workers based on a particular piece of equipment or machinery. The employer must ensure the devices are durable, standardized, and that they do not fail quickly. The procedures must identify how the individual workers will identify themselves on the lock or tag. These systems also need to be reviewed and updated annually. Since most construction sites are multi-employer workplaces, it will be necessary to deconflict all hazardous energy plans prior to beginning work. Lockout procedures that the plan should include, in order:

- Alert the operator(s) that power is being disconnected.
- Preparation for shutdown
- Equipment shutdown
- Equipment isolation
- Application of lockout devices
- Control of stored energy

Control Devices

Devices are used to lockout or tagout the energy flow for a particular machine or operation. Lockout is achieved by using an energy isolating device that can be locked out by a hasp or some other means of attachment to which a lock is affixed. There are other means of isolating devices that are capable of being locked out that can be used if the lockout is achieved without dismantling, rebuilding, or replacing the energy isolating device or permanently alter its control capability.

Devices include manually operated electrical circuit breakers; disconnect switches, line valves, and blocks. It is necessary for the devices and procedures used to lockout equipment are those written in the employer's plan. This ensures that the energy isolating devices used are correct for the equipment used in that company.

Employers can provide the option to tagout the equipment; however, the company must meet specific requirements. First, the placement of a tagout device must provide protection equal to or greater than a lockout device would. Secondly, the tagout device is for situations where a lockout device cannot be used. Lastly, the tagout device must provide a prominent warning and be installed in accordance with the established procedure the company wrote in the plan. It is normal for tagout devices to serve as the written notice to the lockout device to ensure a warning is in place.

Whichever method is used; it is essential to regulate the lockout and tagout devices that are used by workers. It is best to standardize them by color; shape; or size; print and format. Helpful verbiage on tags includes:

- Do Not Start
- Do Not Open
- Do Not Close
- Do Not Operate
- Do Not Energize

Safely removing a lockout or tagout device is just as important as the installation. Only the worker that installed the lockout or tagout device and whose name is on the tag should remove it. If that worker is no longer there, the supervisor can remove if they follow the OSHA exception explained earlier in this chapter. It is important to remember to tell workers before the lockout or tagout device is removed and before the machine or equipment is started.

Training

General and specific training should be provided. The general training addresses the standard and general information about controlling hazardous energy sources. The general training might include:

- Overview
- Scope and Application
- Purpose
- Definition
- Procedures
- Training

OSHA provides an interactive training program on its website (OSHA, 2014). There are also many organizations that provide this type of training, and a simple search of the World Wide Web provides ample choices. The general training can even be provided in a web-based format. It is important to keep a list of the workers that attend and give an examination to show the training took place.

The specific part of the training should be classroom and involve hands-on activities. The specific training is tailored to the written procedures and should include:

- Means of controlling the energy.
- Methods of isolating the energy.
- Means of isolating the energy source.
- How to recognize hazardous energy sources
- How to identify the magnitude of the hazard
- How to identify the type of energy used in the source
- Purpose and function of the energy control program
- Knowledge and skills required for the safe application, usage, and removal of the power control devices.

Hands-on training enables the employer to be sure that each worker knows, understand, and can follow applicable provisions of the written procedures. Workers need to be retrained when:

- They demonstrate a lack of proficiency.
- Changes to the program have occurred.

Summary

Once employers know that workers can die from injuries sustained when working with hazardous energy sources, they must take appropriate steps to prevent those injuries. OSHA has an entire page devoted to this hazard that can be accessed at https://www.osha.gov/SLTC/controlhazardousenergy/. The five main causes of LOTO injuries should be the focus of any program. OSHA Publication 3120 titled "Control of Hazardous Energy (Lockout/Tagout)" is an excellent source for additional information. This publication outlines OSHA's general requirements for controlling hazardous energy during service or maintenance of machines or equipment and is at http://www.osha.gov/Publications/osha3120.pdf.

Chapter 13–Hazard Communication or Global Harmonization

There are several hazardous chemicals and substances that exist in today's modern workplace. Before workers' can be expected to work with these materials safely, the employer must have developed a plan to ensure risks have been identified and controlled. OSHA 29 Code of Federal Regulation, Part 1910, Subpart 1200 titled "Hazard Communication Standard" is the primary source for these hazards. This program is a general industry standard that applies to all industries that include construction (29 CFR, 2014).

The Hazard Communication Standard (HCS) establishes uniform requirements to ensure the hazards of chemicals are evaluated and that the hazard information is provided to employers and exposed workers. For workers who have been in the workforce for a while may also remember this was originally called the "Right to Know Law" (Fanning, 2003). That name describes the program a lot better.

"In March of 2012 OHSA modified the HCS to align it with the United Nations' Globally Harmonized System of Classification and Labeling of Chemicals (GHS). OSHA believes that aligning the HCS with the GHS improves the HCS and helps improve worker safety and health. Data collected and analyzed by OSHA reflects this critical need to develop the HCS. Chemical exposures often result in severe injuries and illnesses among exposed employees. Many workers do not report occupational illnesses because they do not understand that they were exposed at work. This lack of reporting could be because there are long periods of time between exposure and the disease." (Fanning, 2013)

The HCS covers all hazardous chemicals and incorporates a "downstream flow of information." For these standard, hazardous chemicals are one with a physical or health hazard. The producers of chemicals generate and disseminate information. The users of the chemicals obtain that information and make it available to the worker using or exposed to the chemical. Hazardous chemicals are exempt when they are used in isolated instances for Office workers in minuscule quantities and are like chemicals that might be used in the home.

GHS Implementation

The revised HCS affects manufacturers and importers of hazardous chemicals. Chemical manufacturers and importers are required classify chemicals according to the new criteria. Chemical manufacturers and importers prepare and distribute the modified labels and safety data sheets (SDS) to customers. Uniformity is a key benefit of the change that is created by following the detailed rules. These new procedures prevent classifiers reaching different interpretations of the same data (Fanning, 2013).

Threshold Limit Values are required on SDSs. The International Agency for Research on Cancer and the National Toxicology Program classification are also required on SDSs. OSHA finds that requiring this additional information on the SDSs provides useful information to help users assess the hazards. OSHA requires red frames for labels. This frame provides substantial benefit to users because red makes the warnings more visible, and it means the greatest degree of hazard.

"Employers were required to train employees about the new label information and SDS format by December 1, 2013. Employers must comply by June 1, 2015. Companies shall update any alternative workplace labels and provide training for new hazards no later than June 1, 2016." (Fanning, 2013)

"OSHA recognized the need for disclosure of chemical hazard information with the HCS. OSHA estimates that 880,000 hazardous chemicals are used in the United States. The need originally identified in 1983 still exists today. Unfortunately, despite the HCS, workers are exposed to hazardous chemicals." (Fanning, 2013)

Responsibilities

The producers of the chemicals generate and disseminate information about the hazardous chemicals. The producer documents this information using the SDS and labels. The producer must provide customers with a copy of this SDS upon request.

The company determines their responsibilities by doing an inventory of chemicals used on their job sites. Once the inventory is done each label must be read thoroughly to determine if the chemical is hazardous. Employers must develop a written HCS program if they are using hazardous chemicals. The written HCS Program identifies all the other responsibilities the employer has. The second responsibility is to take the inventory taken to determine if the program was needed and eliminate any non-hazardous chemicals and then keep the list as a hazardous chemical inventory. The employer must make that list available to workers if they request to see it.

The next step is to identify procedures for obtaining and maintaining SDS. The company should receive an SDS with each shipment or purchase of hazardous chemicals. Employers must keep SDSs in a central location for workers to access it they choose. The best location is usually in the job trailer at a construction site. If a hazardous chemical comes without an SDS, the employer can call and get one sent to their facility. The employer should have someone periodically check to verify containers of hazardous chemical are labeled. The company ensures that the labels stay on the containers of hazardous chemicals and add labels to any containers that hazardous chemicals are transferred into.

Companies provide training to any worker who will work with or may become exposed to a hazardous chemical on the hazards of the chemical, ways to control the hazards, PPE to wear, and what to do in an emergency.

The worker has several responsibilities too. The worker must attend and apply the training they receive on working with hazardous chemicals. The workers must raise questions if they do not understand any part of the HCS Program. They also must work with a hazardous chemical the way they were trained, which includes wearing PPE. Workers must also report hazards associated with working with hazardous chemicals to their supervisor.

Written Requirements

The written HCS Program is at the heart of effective hazard prevention while working with hazardous chemicals. The written program is a standard operating procedure (SOP) that outlines the entire program. This SOP can then be used to train workers. The written program starts by identifying the responsibilities for the organization. A few are outlined here, but for larger organizations the employer might have responsibilities for a buyer or safety representative that will need to be added. This document will also include the inventory of hazardous chemical as an appendix. The procedures for obtaining and maintaining SDS will be spelled out. Who will obtain SDSs? Who will maintain SDSs? Where will SDSs be maintained? These questions are answered in a written program. The procedures that following container labelling must be spelled out. There must also be a section that addresses training, what the training will include and how often. Including requirements to develop and maintain an attendance roster are necessary to verify those that attend. The employer must review the written plan annually and update it when systems change or if new hazardous chemicals are used. The goal is to ensure procedures are in place so that workers receive the information they need it. Do not forget to identify how workers will be trained on hazards of non-routine tasks and unlabeled pipes when these apply to your work sites.

Labels

The producers are required to put labels on each hazardous chemical they manufacture. Employers need to make sure that each container of hazardous chemicals entering the workplace has a label. That begs the question, what must be on a label? Labels must contain the following information:

- Identity of the chemical
- Hazard warnings

- Name and address of the producer
- The hazard warnings, which can be any type of message, picture, or symbol that provides information on chemical hazards and the targeted organs affected.

Labels must be in English and prominently displayed. There are some exemptions to the labelling requirements. Portable containers that are used to transfer hazardous chemicals into are not required to be labeled containers if the container is intended only for the immediate use of the worker who transferred the chemical and only for the current shift.

Safety Data Sheets (SDS)

Producers and importers provide an SDS for each hazardous chemical they manufacture and import. SDS used to be called Material Safety Data Sheets or MSDSs. Employers need to make sure that each container of hazardous chemicals in the workplace has an SDS. It is important to know that each container shipped might mean a pallet that will have one SDS. Copies will have to be made of that SDS when the pallet is broken down and the containers are sent to each job site. Electronic access to an SDS is acceptable if the employer has a backup procedure in place. The following information must be on an SDS:

- Name, address, and telephone number of producers
- Specific chemical identity and common names
- Routes of exposure and control measures
- Physical hazards and Exposure Limits
- Precautions for safe handling and use
- Physical and chemical characteristics
- Emergency and first-aid procedures
- Health hazards and effects
- Carcinogenicity

There may still be MSDSs in use and some may be rather old. That is because the MSDS or SDS is only updated within three months of new or significantly changed information about the chemical's hazard potential. There is no requirement to update them if nothing changes. That means that an old MSDS could be correct.

Training

Employers are required to provide training for workers exposed to or may become exposed to hazardous chemicals in their work. A best practice is to do the training when assigning the worker and when introducing new hazards into their work.

Training must ensure that the worker knows how to work with a hazardous chemical correctly and safely. All this information should be in the HCS program. The company can use one of several methods that include computer-based training, webinars, classroom instruction, or interactive videos. With these methods, the employer should always have a trainer on hand to answer any questions workers may have. Make sure everyone signs-in and is given an examination in the end of training to verify that the workers learned the material.

It is okay to provide general training on the HCS program, but the training must cover each hazardous chemical found in the workplace. If the employer has workers that are trained by a former employer or union, the employer does not have to retrain them if the previous training met the standards for the current work. Of course, this means the company will still need to tell the workers where to find SDSs at their work site, who are responsible for the HCS program, and where the worker can see the written program.

Summary

A simple concept is the basis for the OSHA's Hazard Communication Standard – employees have a right to know about the hazards they work with. OSHA has an entire page dedicated to these hazards that can be accessed at https://www.osha.gov/dsg/hazcom/solutions.html.

Summary

The manager or owner of a construction company sets the vision of what the safety program is and the goals that it has. The company develops policies that direct the safety program and more importantly ensures funding is available for the program. The company may also reward personnel for safety performance.

The supervisor is the person from management most frequently in contact with the worker. In this capacity, he or she often is the role model. The supervisor must communicate management's policies and standards to the work force. He or she must then follow-up by enforcing these policies through their actions as well as correcting workers when there are in violations.

Safety must be considered in project development as well as equipment and personnel selection. Everyone needs to perform in accordance with the training and directions provided by his or her supervisor and to report unsafe and unhealthful working conditions, as well as other workers acting unsafe.

The focus should be on behavior. Supervisors need to identify the Activators present and the Competencies demonstrated that lead to behavior. Supervisors then provide consequences in response to the action. It is difficult to change a person's thoughts, values, and beliefs. It is not difficult to change a worker's behavior. That is because behaviors can be seen and therefore are the best thing to emphasize. Taking the information in this book and ensuring workers are demonstrating the correct behavior can save a lot of resources.

Fred E. Fanning

#####

If you would like to help other readers out, please leave a review of this book on Amazon.com. Your rating and review will help them decide to read or not read this book.

#####

From my other books, I recommend Safety Risk Management. You can see it at the following URL https://www.amazon.com/gp/product/B01IVX5UTS/ref=dbs_a_def_rwt_bibl_vppi_i14.

Bibliography

- 29 Code of Federal Regulations, Part 1910, Subpart 1200, Hazard Communication retrieved on May 20, 2014 from https://www.osha.gov/pls/oshaweb/owadisp.show_document?p_table=STANDARDS&p_id=10099.
- 29 Code of Federal Regulation, Subpart M, 1926.501 retrieved on January 25, 2014 from https://www.osha.gov/pls/oshaweb/owadisp.show_document?p_table=STANDARDS&p_id=10757
- 29 Code of Federal Regulation, Subpart M, 1926.502 retrieved on January 25, 2014 retrieved from https://www.osha.gov/pls/oshaweb/owadisp.show_document?p_id=10758&p_table=STANDARDS
- 29 Code of Federal Regulation, Subpart K: Electrical, 1926, retrieved on February 14, 2014 from https://www.osha.gov/pls/oshaweb/owadisp.show_document?p_table=STANDARDS&p_id=10915.
- A Guide to Scaffold Use in the Construction Industry, OSHA Booklet Number 3150 2002 (Revised) Retrieved on January 19, 2014 from https://www.osha.gov/Publications/osha3150.pdf.
- Accident Prevention Signs and Tags, 29 CFR, 1926.200. Retrieved on March 11, 2014 from: https://www.osha.gov/doc/outreachtraining/htmlfiles/subpartg.html.
- American Ladder Institute, Choosing the Right Ladder. Retrieved on July 18, 2014 from http://www.americanladderinstitute.org/?page=ChoosingaLadder.
- Construction Industry retrieved on November 30, 2013 from https://www.osha.gov/doc/index.html
- Competent Person retrieved on April 22, 2014 from https://www.osha.gov/SLTC/competentperson/index.html.
- Della-Giustina, Daniel E., Developing a Safety and Health Program, Lewis Publishing, 2000.
- EHS Works "Preventing Backing Incidents on Construction Sites" retrieved on August 11, 2014 from http://ehsworks1.blogspot.com/2014/08/preventing-backing-incidents-on.html.

- Exception to 29 CFR 1910.147, (e), (3) retrieved on April 28, 2014 from https://www.osha.gov/pls/oshaweb/owadisp.show_document?p_id=9 804&p_table=STANDARDS.
- Fanning, Fred, "Basic Safety Administration: A Handbook for the New Safety Specialist." American Society of Safety Engineers, Revised edition, June 2003, USA.
- Fanning Fred, GHS in Summary, American Society of Safety Engineers, Northern Virginia Chapter Newsletter, First Quarter NOVA News, Issue 1 February 2013 retrieved on May 19, 2014 from http://nova.asse.org/wp-content/uploads/2013/02/NOVANewsletter_2013Q1.pdf.
- Federal Register, Proposed Construction Confined Space Rule, retrieved on May 16, 2014 from https://www.osha.gov/pls/oshaweb/owadisp.show_document?p_table =FEDERAL_REGISTER&p_id=20174.
- Forklift Training, 29 CFR 1910.178, Powered industrial trucks. Retrieved on March 11, 2014 from: https://www.osha.gov/pls/oshaweb/owadisp.show_document?p_table =STANDARDS&p_id=9828.
- Four Causes of Construction Accidents retrieved on October 17, 2013 from https://www.osha.gov/Publications/3216-6N-06-english-06-27-2007.html.
- Legal Definitions retrieved on October 17, 2013 from http://definitions.uslegal.com/n/national-consensus-standard/.
- Letters of Interpretation retrieved on October 17, 2013 from https://www.osha.gov/pls/oshaweb/owadisp.show_document?p_table =INTERPRETATIONS&p_id=25178.
- Motor Vehicles, 29 Code of Federal Regulation, Part 1926, Subpart O–Motor Vehicles, Mechanized Equipment, and Marine Operations. Retrieved on March 11, 2014 from: https://www.osha.gov/pls/oshaweb/owadisp.show_document?p_table =STANDARDS&p_id=10929.
- Moran, Mark, Construction Safety Handbook 2nd ed, published by Government Institutes, Inc., 2003.
- OSHA Fall Protection Page retrieved on January 25, 2014 from https://www.osha.gov/SLTC/fallprotection/.
- National Electric Code, NFPA 70, published by the National Fire Protection Association, 2013.
- OSHA Scaffolding Page retrieved on February 1, 2014 from https://www.osha.gov/doc/outreachtraining/htmlfiles/scaffreg.html.

- OSHA 19CFR 1926, Subpart P App B retrieved on April 22, 2104 from https://www.osha.gov/pls/oshaweb/owadisp.show_document?p_table=STANDARDS&p_id=10932.
- OSHA Booklet 2226 retrieved on April 22, 2014 from https://www.osha.gov/Publications/OSHA2226/2226.html.
OSHA training website retrieved on April 28, 2014 from https://www.osha.gov/dts/osta/lototraining/index.html.
- OSHA Booklet 3111, Hazard Communication Guidelines for Compliance retrieved on May 19, 2014 from http://www.osha.gov/Publications/osha3111.pdf.
- Publication 3120 "Control of Hazardous Energy (Lockout/Tagout)" retrieved on April 28, 2014 from http://www.osha.gov/Publications/osha3120.pdf.
- ROPS, 29 CFR1926 subpart W - Rollover Protective Structures; Overhead Protection. Retrieved on March 11, 2014 from https://www.osha.gov/pls/oshaweb/owadisp.show_document?p_table=STANDARDS&p_id=10945.
- Safety and Health Topics, Confined Space Entry, retrieved on May 16, 2014 from https://www.osha.gov/SLTC/confinedspaces/construction.html.
- Signs, 29 Code of Federal Regulation, Part 1926, Subpart G: Signs, Signals and Barricades. Retrieve on March 11, 2014 from: https://www.osha.gov/pls/oshaweb/owadisp.show_document?p_table=STANDARDS&p_id=10911.
- The OSH Act of 1970 retrieved on November 23, 2013 from https://www.osha.gov/pls/oshaweb/owasrch.search_form?p_doc_type=OSHACT&p_toc_level=0&p_keyvalue=&p_status=CURRENT.

About the Author

After a successful career as a Federal Employee that included over twenty years in safety and occupational health. I started writing part-time. My published work includes the peer-reviewed book Basic Safety Administration: A Handbook for the New Safety Specialist in its second edition. I also authored two editions of the peer-reviewed chapter, Safety Training and Documentation Principles published in the bestselling, Safety Professional Handbook, and the Safety Professional Handbook Management Applications, both edited by Joel Haight, Ph.D., CSP. I co-authored the peer-reviewed chapter Safety Training with Christine Fiori, Ph.D., PE, published in the bestselling Construction Safety Management and Engineering, second edition edited by Darryl C. Hill, Ph.D., CSP. The American Society of Safety Professionals Traditionally published my book and chapters.

I self-published another eleven books using Kindle Direct Publishing. Seven of these books are available in paperback and Kindle formats. Four of those books are available only in Kindle format. I have authored over fifty articles in various publications on safety and occupational health and project management. I have earned several writing awards for my non-fiction work and one for my fiction work. I have self-published two novels, A Walk Among the Dead and my most recent Mystery at Devil's Elbow.

I am an Emeritus Professional Member of the American Society of Safety Professionals. I was selected as the Safety Professional of the Year for the Northern Virginia Chapter of this Society. I am also a member of the Non-Fiction Writers Association. I held the Certified Safety Professional (CSP) designation for ten years. I also earned master's degrees from National-Louis University and Webster University.

Index

Fred E. Fanning

Hazard Communication 8, 85, and 90

Hazards 1, 2, 3, 4, 5, 8, 12, 15, 16, 17, 18, 22, 23, 24, 25, 29, 30, 31, 32, 34, 35, 36, 37, 38, 39, 40, 41, 42, 43, 44, 45, 46, 47, 49, 50, 51, 52, 53, 54, 55, 56, 58, 60, 61, 63, 65, 66, 67, 68, 69, 70, 71, 72, 73, 74, 75, 76, 77, 78, 79, 80, 85, 86, 87, 88, 89, and 90

Head Protection 11, 17, and 90

Illness 2, 3, 4, 9, 14, 66, 85, and 90

Injury 9, 22, 28, 31, 33, 35, 43, 45, and 67

Ladder 1, 11, 45, 50, 61, 63, 67, 68, 69, 70, 71, and 77

Lockout 10, 16, 48, 49, 78, 80, 81, 82, 83, and 84

Material Safety Data Sheet (MSDS) 89

Mechanized Equipment 9, 24, and 25

Motor Vehicle 1, 11, 24, 25, 28, and 63

National Electric Code 7 and 44

Occupational Safety and Health Administration (OSHA) 1, 2, 3, 4, 5, 6, 7, 8, 9, 10, 12, 13, 14, 15, 16, 19, 23, 24, 30, 38, 49, 52, 55, 59, 62, 63, 64, 65, 66, 71, 80, 81, 83, 84, 85, 86, and 90

Permits 10, 42, 43, 47, 72, 74, 75, and 78

Personal Fall Arrest Systems (PFAS) 51, 53, 57, and 58

Personal Protective Equipment (PPE) 1, 15, 16, 19, 21, 23, 31, 34, 35, 37, 38, 39, 47, 66, 76, 78, and 87

Power Tools 31, 35, 36, 37, 38, and 47

Safety 1, 2, 3, 4, 5, 7, 8, 10, 11, 12, 15, 16, 20, 23, 30, 38, 41, 42, 49, 54, 56, 57, 58, 65, 68, 70, 72, 74, 75, 76, 81, 85, 86, 88, 89, and 91

Appendix A - Confined Space Entry

Item	Question	Yes	No
1	Is there a written document that outlines this program?		
2	Is this document reviewed annually and updated when changes occur?		
3	Are all operations reviewed to identify confined spaces?		
4	Is a list made of all confined spaces?		
5	Is a list made of all permit required confined spaces?		
6	Are procedures in place to provide for emergency rescue and treatment of personnel injured in a confined space operation?		
7	Are employees trained on the specific duties they hold during a confined space operation?		
8	Are supervisors trained on their duties?		
9	Are rescue teams appointed and trained?		
10	Is the proper Personal Protective Equipment (PPE) provided to employees?		
11	Are supervisors provided adequate test equipment to test for hazards within the confined space?		
12	Are the requirements for obtaining a permit spelled out in the program document?		

Appendix B - Control of Energy Sources

Item	Question	Yes	No
1	Is there a written document that outlines this program?		
2	Is this document reviewed annually and updated when changes occur?		
3	Are all operations reviewed to identify processes where the energy source must be controlled during maintenance operations?		
4	Are procedures in place to provide for tagging or locking out of the energy source before maintenance begins?		
5	Are employees trained on the specific duties they hold during maintenance operations, to include the importance of not energizing a machine that has been tagged or locked-out?		
6	Are supervisors trained on their duties?		
7	Is the proper tagging and lockout devices made available to Employees?		
8	Is there an evaluation program in place to check this program for effectiveness?		
9	Are employees retrained when equipment is changed, or they demonstrate deficient procedures?		

Appendix C part 1 of 2 - Hazard Communication

Item	Question	Yes	No
1	Does each element within your organization have a complete list of hazardous substances used or stored within the workplace?		
2	Is the list maintained in the workplace?		
3	Is the list made available to employees of the work area upon their request?		
4	Is the list updated when hazardous substances are added or subtracted from the workplace?		
5	Are all containers of hazardous substances labeled?		
6	Are containers without labels set aside and not used until the proper label can be attached to the container?		
7	Are secondary containers properly labeled after substances are poured into them from primary containers?		
8	Are Material Safety Data Sheets (MSDSs) available on all hazardous substances in the workplace?		
9	Are the MSDSs maintained in the workplace?		
10	Are the MSDSs provided to employees of the workplace upon request?		
11	Are MSDS added when a new hazardous substance is added to the workplace?		

Appendix C part 2 of 2 - Hazard Communication

Item	Question	Yes	No
12	Are less hazardous substances substituted for more dangerous ones whenever possible?		
13	Are employees who may encounter a hazardous substance provided with training on their rights and obligations under the Hazard Communication Program?		
14	Do employees receive specific training on the hazards of a particular material they are working with as well as the proper method of working with the substance?		
15	Is Personal Protective Equipment provided to employees to control exposure to hazards?		
16	Do employees receive training on the use and maintenance of the PPE?		
17	Does the organization have a written document that addresses all elements of this program?		
18	Is this document or standard operating procedure reviewed annually and updated when changes occur?		
19	Does the employer have an evaluation program in place to check this program?		
20	Are the new GHS requirements integrated into the implementation?		

Appendix D - Motor Vehicle Accidents

Item	Question	Yes	No
1	Is there a written document that outlines this program?		
2	Is this document reviewed annually and updated when changes occur?		
3	Are potential drivers passing a physical?		
4	Are drivers professionally trained and licensed on the vehicle they will be hired to operate?		
5	Do employees complete a defensive driving type course?		
6	Do supervisors evaluate driver's performance?		
7	Are drivers with traffic violations or accidents counseled?		
8	Are drivers retrained when a training weakness is identified and not as punishment?		
9	Is an awards program in place to reward good and safe drivers?		
10	Is an awareness program in place to raise driver awareness for specific hazards?		
11	Is there an evaluation conducted by management to evaluate the effectiveness of this program?		

Appendix E - General Personal Protection

Item	Question	Yes	No
General			
1	Is there a written document that outlines this program?		
2	Is this document reviewed annually and updated when changes occur?		
3	Have all facilities and operations been inspected?		
4	Have hazards that need an engineering correction been identified and preparations made for correction?		
5	Have engineering corrections that require more than a few days to implement been identified for temporary personal protective equipment (PPE) abatement?		
6	Have hazards requiring permanent resolution by PPE been identified?		
7	Have specific items of PPE been identified to reduce or eliminate the hazard?		
8	Have procedures been developed to ensure replacement items of PPE are available for workers?		

Appendix F part 1 of 2 - Eye Protection

Item	Question	Yes	No
Impact Protection			
1	Have hazards been identified that require protection for the eyes from projectiles?		
2	Have impact protection goggles or glasses been identified for use to protect the eye against the projectiles?		
3	Have styles and sizes that promote proper wear been purchased?		
4	Have the goggles or glasses been fitted correctly to the worker?		
5	Has the worker been trained in the proper maintenance and cleaning of the goggles and glasses?		
Chemical or Vapor Protection			
1	Have hazards been identified that require protection from splashing chemicals or vapors?		
2	Have goggles or glasses been identified for use to protect the eye against the splashing chemicals or chemical vapors?		
3	Have styles and sizes that promote proper wear been purchased?		
4	Have the goggles or glasses been fitted correctly to the worker?		

Appendix F part 2 of 2 - Eye Protection

Item	Question	Yes	No
Arc Welding Protection			
1	Have risks been identified for protection of the eyes against the hazards of Arc Welding?		
2	Has a style and size that promotes proper wear been purchased?		
3	Has a protective hood been fitted correctly to the worker?		
4	Has the worker been trained in the proper maintenance and cleaning of the hood?		
5	Has the worker also been fitted and issued a proper pair of impact protection goggles or glasses to protect against the slag removal? See the section on impact protection for specifics		
Gas Welding Protection			
1	Have risks been identified for protection of the eyes from the hazards of Gas Welding?		
2	Have proper protective goggles or glasses been identified for use to protect the eye against the light?		
3	Have styles and sizes that promote proper wear been purchased?		
4	Have the goggles or glasses been fitted correctly to the worker?		

Appendix G - Head and Foot Protection

Item	Question	Yes	No
Head Protection			
1	Have risks been identified that require protection against the hazards of objects falling or being dropped onto the head?		
2	Has a proper protective helmet or hat been identified for use to protect the head against the falling or dropped object?		
3	Has a style and size that promote proper wear been purchased?		
4	Has the helmet or hat been fitted correctly to the worker?		
5	Has the worker been trained in the proper maintenance and cleaning of the helmet or hood?		
Foot Protection			
1	Have risks been identified that require protection against the hazards of objects falling or being dropped onto the feet?		
2	Has a proper shoe or guard been identified for use to protect the feet against the falling or dropped object?		
3	Has a style and size that promote proper wear been purchased?		
4	Has the safety shoe or guard been fitted correctly to the worker?		

Appendix H - Hand and Body Protection

Item	Question	Yes	No
Hand Protection			
1	Have hazards been identified that require protection from cold and hot temperatures, abrasions, cuts, vibration, or chemical exposure to the hands?		
2	Has a proper pair of gloves been identified for use to protect the hand against these hazards?		
3	Has a style and size that promote proper wear been purchased?		
4	Have the gloves been fitted correctly to the worker?		
5	Has the worker been trained in the proper maintenance and cleaning of the gloves?		
Body Protection			
1	Have hazards been identified that require protection for the body from extreme cold and hot temperatures?		
2	Has proper clothing (pants, jacket, gloves, and hood) been identified for use to protect the body from these hazards?		
3	Has a style and size that promote proper wear been purchased?		
4	Has the clothing been fitted correctly to the worker?		

Appendix I - Ear Protection

Item	Question	Yes	No
Noise Protection			
1	Have hazards been identified that require protection for the ears from extreme noise?		
2	Has proper earmuffs or plugs been identified for use to protect the ear from noise?		
3	Has a style and size that promote proper wear been purchased?		
4	Have the muffs or plugs been fitted correctly to the worker?		
5	Are noise reduction ratings used to determine which hearing protection devices to use?		
6	Are ear plugs provided for visitors?		
7	Has the worker been trained in the proper maintenance and cleaning of the muffs or plugs?		
8	Are employees with an eight-hour average noise exposure of 85 decibels monitored?		
9	If an employee is on the monitoring program do, they receive annual audiometric testing?		
10	If an employee experiences, loss of hearing is it reported to OSHA?		

Appendix J - Respiratory Protection

Item	Question	Yes	No
Respiratory Protection			
1	Have risks been identified that require protection for the respiratory tract (lungs, throat, and sinuses) from hazards? Note: These hazards should be identified and quantified by an Industrial Hygienist or a Certified Safety Professional.		
2	Is there a respiratory protection program within the organization?		
3	Has the proper respirator been identified for use to protect the respiratory tract? Note: An Industrial Hygienist should help select the respirator.		
4	Has a style and size that promote proper wear been purchased?		
5	Have workers passed a physical certifying him or her to wear a respirator?		
6	Has the respirator been fit tested specifically to the worker?		
7	Has the respirator fit test been conducted by a person certified to fit test?		

Appendix K - Fire Prevention and Protection

Item	Question	Yes	No
1	Has the need and location for all fire extinguishers been determined?		
2	Have the proper fire extinguishers been purchased and positioned where they are required?		
3	Have employees been trained on where fire extinguishers are located and how to operate them?		
4	Is there a plan in place to maintain and refill fire extinguishers?		
5	Have the proper amount and type of fire exits been designated?		
6	Are flammable substances stored in flameproof cabinets or rooms?		
7	Are ignition sources prohibited near combustible or flammable substances?		
8	Are smoking areas kept to a minimum and located away from combustible or flammable sources?		
9	Are electrical wiring and tools inspected periodically to prevent electrical shorts?		
10	Have arrangements been made for fire protection and firefighting services to respond to your organization in case of an emergency?		
11	Have arrangements been made for emergency medical services to respond to your organization in case of fire?		

Appendix L part 1 of 3 - Hazard Prevention

Subject	Requirement
Housekeeping	Conditions are orderly
	Floors, aisles, and work areas are clean and dry
	Trash and containers are covered
	Cleaning materials are provided
Portable Ladders	Are in good physical condition
	Have a nonskid base on legs
	Are not left exposed to weather
	Have all rungs in place
	Periodic inspection of ladders is conducted.
	Users are trained to correctly use ladder.
Flammable Liquids	Only approved containers and cabinets are used.
	Combustible waste is in covered containers.
Fire Prevention	Fire extinguishers are accessible and properly maintained.
	Fire extinguishers are the correct type for the hazard.

Appendix L part 2 of 3 - Hazard Prevention

Subject	Requirement
Fire Prevention (continued)	Fire extinguishers are inspected and weighed at least yearly.
Electrical	Main disconnect switches are legibly marked.
	Fixed and portable equipment is grounded.
	Flexible cords are used only as a temporary measure.
	All electrical outlets and switches are in good working order.
Material Handling	Fork-lifts are correct for the hazards of the location they are used in.
	Fork-lifts are in good working order.
	Operators are trained and licensed.
	Forks are lowered when left unattended.
	Fork-lifts have the key removed when operator is not present.
	Fork-lift charging stations for battery filling have an eye station nearby.

Appendix L part 3 of 3 - Hazard Prevention

Subject	Requirement
Compressed Gases	Inspected on receipt and prior to cylinders use.
	Stored in a safe location.
	Stored upright and fastened to a fixed object so they will not fall over.
	Separated by gas type.
	A safety relief valve is installed when necessary.
	If stored outside a cover is present to prevent sunlight from falling on tanks.
Control of Energy Sources	Operating procedures in place to control energy sources during maintenance operations.
	Train employees on the use of tags and lockouts.
	Tags and lockouts are used to control energy sources during maintenance.
	Only personnel who install tags and lockouts remove them. Note: OSHA exception can be used.